Creating Autonomous Vehicle Systems

Second Edition

Synthesis Lectures on Computer Science

The Synthesis Lectures on Computer Science publishes 75–150 page publications on general computer science topics that may appeal to researchers and practitioners in a variety of areas within computer science.

Creating Autonomous Vehicle Systems, Second Edition
Shaoshan Liu, Liyun Li, Jie Tang, Shuang Wu, and Jean-Luc Gaudiot
September 2020

Blockchain Platforms: A Look at the Underbelly of Distributed Platforms
Stijn Van Hijfte
July 2020

Analytical Performance Modeling for Computer Systems, Third Edition
Y.C. Tay
July 2018

Creating Autonomous Vehicle Systems
Shaoshan Liu, Liyun Li, Jie Tang, Shuang Wu, and Jean-Luc Gaudiot
October 2017

Introduction to Logic, Third Edition
Michael Genesereth and Eric J. Kao
November 2016

Analytical Performance Modeling for Computer Systems, Second Edition
Y.C. Tay
October 2013

Introduction to Logic, Second Editio
Michael Genesereth and Eric Kao
August 2013

Introduction to Logic
Michael Genesereth and Eric Kao
January 2013

Creating Autonomous Vehicle Systems, Second Edition
Shaoshan Liu, Liyun Li, Jie Tang, Shuang Wu, and Jean-Luc Gaudiot

ISBN: 978-3-031-00677-7   paperback
ISBN: 978-3-031-01805-3   ebook
ISBN: 978-3-031-00073-7   hard

DOI: 10.1007/978-3-031-01805-3

A Publication in the Springer series
SYNTHESIS LECTURES ON COMPUTER SCIENCE, #12

Series ISSN: 1932-1228 Print   1932-1686 Electronic

Creating Autonomous Vehicle Systems

Second Edition

Shaoshan Liu

PerceptIn

Liyun Li

Xpeng Motors

Jie Tang

South China University of Technology

Shuang Wu

YiTu

Jean-Luc Gaudiot

University of California, Irvine

SYNTHESIS LECTURES ON COMPUTER SCIENCE #12

ABSTRACT

This book is one of the first technical overviews of autonomous vehicles written for a general computing and engineering audience. The authors share their practical experiences designing autonomous vehicle systems. These systems are complex, consisting of three major subsystems: (1) algorithms for localization, perception, and planning and control; (2) client systems, such as the robotics operating system and hardware platform; and (3) the cloud platform, which includes data storage, simulation, high-definition (HD) mapping, and deep learning model training. The algorithm subsystem extracts meaningful information from sensor raw data to understand its environment and make decisions as to its future actions. The client subsystem integrates these algorithms to meet real-time and reliability requirements. The cloud platform provides offline computing and storage capabilities for autonomous vehicles. Using the cloud platform, new algorithms can be tested so as to update the HD map—in addition to training better recognition, tracking, and decision models.

Since the first edition of this book was released, many universities have adopted it in their autonomous driving classes, and the authors received many helpful comments and feedback from readers. Based on this, the second edition was improved by extending and rewriting multiple chapters and adding two commercial test case studies. In addition, a new section entitled "Teaching and Learning from this Book" was added to help instructors better utilize this book in their classes. The second edition captures the latest advances in autonomous driving and that it also presents usable real-world case studies to help readers better understand how to utilize their lessons in commercial autonomous driving projects.

This book should be useful to students, researchers, and practitioners alike. Whether you are an undergraduate or a graduate student interested in autonomous driving, you will find herein a comprehensive overview of the whole autonomous vehicle technology stack. If you are an autonomous driving practitioner, the many practical techniques introduced in this book will be of interest to you. Researchers will also find extensive references for an effective, deeper exploration of the various technologies.

KEYWORDS

autonomous driving, driverless cars, perception, vehicle localization, planning and control, autonomous driving hardware platform, autonomous driving cloud infrastructures, low-speed autonomous vehicle, autonomous last-mile delivery vehicle

Contents

Preface to the Second Edition

Autonomous vehicles—be they on land, on water, or in the air—are upon us and are finding a myriad of new applications, from driverless taxi services to automatic airborne surveillance of sensitive remote areas. Continued technological advances in the past few decades have made these innovations possible, but the design problems that must be surmounted in order to provide useful, efficient, and, supremely importantly, safe operations of these independent units are equally numerous and daunting.

It is thus the purpose of this book to provide an overview of these problems and lead the reader through some common design solutions. High technological capabilities, complete integration of hardware and software, and deep synergy with resident platforms (such as cloud servers) are a must for an eventual successful deployment. The focus of the book is on land vehicles and, more specifically, cars in urban or country road environments, as well as off-road operations. The aim of this book is to address an audience of engineers, be they from the academic or the industrial side, with a survey of the problems, solutions, and future research issues they will encounter in the development of autonomous vehicles, from sensing, perception, all the way to action, and including support from cloud-based servers. A copious amount of bibliographic references completes the picture and will help the reader navigate through a jungle of past work.

The first edition of this book was very well received, as many universities adopted it in their autonomous driving classes, and many companies utilized it for their internal trainings (more details on this can be found in the new chapter "Teaching and Learning from this Book"). Also, we are very fortunate to have received a lot of helpful comments and feedbacks from the first edition readers, which greatly helped improve the content and the quality of the second edition. Specifically, based on these comments and feedback, we made the following improvements in the second edition. First, we extended Chapters 3 and 4 to provide the latest updates on perception techniques; second, we rewrote Chapters 5, 6, and 7 to make them easier to understand; third, we added two commercial case studies in Chapters 10 and 11, so that readers can easily understand how to apply their learnings in real-world environments. In addition, we added a section titled "Teaching and Learning from this Book" to help instructors better utilize this book in their classes. We believe that the second edition captures the up-to-date advances in autonomous driving, and also presents practical real-world case studies to help readers utilize their learnings in commercial autonomous driving projects.

STRUCTURE OF THE BOOK

A brief history of information technology and an overview of the algorithms behind autonomous driving systems, of the architecture of the systems, and of the support infrastructure needed is provided in Chapter 1. Localization, being one of the most important tasks in autonomous driving, is covered in Chapter 2, where the most common approaches are introduced. The principles, advantages, and drawbacks of GNSS, INS, LiDAR, and wheel odometry are described in detail, and the integration of various versions of these strategies are discussed. As for detection, i.e., "understanding" the environment based on sensory data, it is described in Chapter 3, with an exploration of the various algorithms in use, including scene understanding, image flow, tracking, etc. The large datasets, highly complex computations required by image classification, object detection, semantic segmentation, etc., are best handled by the deep learning approaches to perception advocated for in Chapter 4, where applications to detection, semantic segmentation, and image flow are described in detail. Once the environment is understood by the autonomous vehicle, it must somehow predict future events (e.g., the motion of another vehicle in its vicinity) and plan its own route; this is the purpose of Chapter 5. Next, in Chapter 6, an even more detailed level of decision making, planning, and control is presented. Feedback between modules with possibly orthogonal decisions as well as conflict resolution (e.g., one module could recommend a lane change, but another one has detected an obstacle in the lane in question) are covered with an emphasis on describing algorithms for behavioral decision making (e.g., Markov decision processes, scenario-based divide and conquer), and for motion planning. This leads into Chapter 7 for a demonstration of the need to supplement the design with Reinforcement Learning-based Planning and Control for a complete integration of situational scenarios in the development of an autonomous system. Underneath it all, the on-board computing platform is the topic of Chapter 8. It includes an introductory description of the Robot Operating System, followed by an actual summary of the real hardware employed. The need for heterogeneous computing is introduced with a strong emphasis on meeting real-time computing requirements as well as on-board considerations (power consumption and heat dissipation). This means that a variety of processing units (general-purpose CPU, GPUs, FPGAs, etc.) must be used. Chapter 9 covers the infrastructure for the cloud platform used to "tie it all together" (i.e., provide services for distributed simulation tests for new algorithm deployment, offline deep learning model training, and High-Definition (HD) map generation). Chapter 10 presents a case study of a commercial autonomous last-mile delivery vehicle operating in complex traffic environments. Finally, Chapter 11 presents a case study of affordable autonomous vehicles for microtransit services.

Teaching and Learning from This Book

1. INTRODUCTION

In recent years, autonomous driving has become quite a popular topic in the research community as well as in industry. However, the biggest barrier to the rapid development of this field is a very limited talent supply. This is due to several problems: first, autonomous driving is the complex integration of many technologies, making it extremely challenging to teach; second, most existing autonomous driving classes focus on one technology of the complex autonomous driving technology stack, thus failing to provide a comprehensive introduction; third, without good integration experiments, it is very difficult for the students to understand the interaction between different technology pieces.

To address these problems, we have developed a modular and integrated approach to teach autonomous driving. For students interested in autonomous driving, this book provides a comprehensive overview of the whole autonomous vehicle technology stack. For practitioners, this book presents many practical techniques and a number of references to aid them performing an effective, deeper exploration of particular modules. In addition, to help the students understand the interaction between different modules, we developed platforms for hands-on integration experiments. Our teaching methodology starts with an overview of autonomous driving technologies, followed by different technology modules, and ends with integration experiments. Note that the order of the modules can be flexibly adjusted based on the students' background and interest level. We have successfully applied this methodology to three different scenarios: an introduction to an autonomous driving class for undergraduate students with limited technology background; a graduate level embedded systems class, in which we added a session on autonomous driving; as well as a two-week professional training for seasoned engineers.

The rest of this overview is organized as follows. Section 2 reviews existing classes on autonomous driving; Section 3 presents the details of the proposed modular and integrated teaching methodology; Section 4 presents the three case studies where we applied the proposed methodology; and we draw the conclusions in Section 5.

2. EXISTING CLASSES ON AUTONOMOUS DRIVING

Autonomous driving has been attracting attention from academia as well as industry. However, comprehensive and complex autonomous driving systems involve a very diverse set of technologies including sensing, perception, localization, decision making, real-time operating system, heterogeneous computing, graphic/video processing, cloud computing, etc. This sets high requirements for lecturers to master all aspects of relative technologies. It is even more of a challenge for students to understand the interactions between these technologies.

Problem-based learning (PBL) is a feasible and practical way to teach relative knowledge and technologies of autonomous driving [20]. Costa et al. built a simulator by integrating Gazebo 3D simulator for students to design and understand autonomous driving systems [13]. In terms of the educational methodology on Learning from Demonstration, Arnaldi et al. proposed an affordable setup for machine learning applications of autonomous driving by implementing embedded programming for small autonomous driving cars [14]. However, these approaches usually cover only one or two technologies, such as machine learning, and fail to provide a comprehensive understanding of the whole system.

Several major universities already offer autonomous-driving-related classes. For instance, MIT offers two courses on autonomous driving. The first one focuses on Artificial Intelligence and is available online to the public and registered students. This course invites several guest speakers on the topic of deep learning, reinforcement learning, robotics, psychology, etc. [15]. The other concentrates on deep learning for self-driving cars and teaches deep learning knowledge by building a self-driving car [16]. Stanford also offers a course for introducing the key artificial intelligence technologies that could be used for autonomous driving [17]. However, these classes all focus on machine learning and do not provide good coverage of the different technologies involved in autonomous driving. It is thus hard for students to obtain a comprehensive understanding of autonomous driving systems.

In terms of experimental platforms and competition developments, Paull et al. proposed Duckietown, an open-source and inexpensive platform for autonomy education and research [18]. The autonomous vehicles are equipped with a Raspberry Pi 2 and a monocular camera for sensing. In addition, there are several competitions of autonomous driving for a broad range of students to stimulate their passion and encourage learning about key technologies in autonomous cars and related traffic systems [19, 21–23]. However, to use these platforms and to enter these competitions, students need to first gain a basic understanding of the technologies involved as well as their interactions; this type of education is currently lacking.

As seasoned autonomous driving researchers and practitioners, we think the best way to learn how to create autonomous vehicle systems is to first grasp the basic concepts in each technology module, then integrate these modules to understand how they interact. Existing autonomous driving classes either focus on only one or two technologies, or have the students directly build a working autonomous vehicle. As a result, this disconnection between individual technology module and system integration creates a high entry barrier for students interested in autonomous driving and often intimidates interested students away from entering this exciting field. To address this exact problem, we designed our modular and integrated approach, which we will present in this chapter, along with sharing our experiences with autonomous driving education.

3. A MODULAR AND INTEGRATED TEACHING APPROACH

Autonomous driving involves many technologies, thus making autonomous driving education extremely challenging. To address this problem, we propose a modular and integrated teaching methodology for autonomous driving. To implement this methodology, we have developed modular teaching materials including a textbook and a series of multimedia online lectures, as well as integration experimental platforms to enable a comprehensive autonomous driving education.

3.1 TEACHING METHODOLOGY

In the past several years, we have taught undergraduate- and graduate-level classes as well as initiated new engineers to the concepts of autonomous driving. A common problem that we found was that the first encounter on this subject usually intimidated many students because of its perceived complexity. Similarly, even experienced engineers newly exposed to the field of autonomous driving felt it was extremely stressful (in other words, it took them outside their comfort zone), especially since the subject touches upon many new areas.

On the other hand, we found that a modular and integrated approach is an effective way of teaching autonomous driving. This means that we first break the complex autonomous driving technology stack into modules and have the students start with the module with which they are most familiar, then have them move on to other modules. This allows the students to maintain a high level of interest and make satisfactory progress throughout the learning process. Once the students have gone through all the modules, they are challenged to perform a few integration experiments to understand the interactions between these modules. Aside from being an effective teaching method, this approach allows the instructors to flexibly adapt the class curriculum to the needs of students with different technology backgrounds, including undergraduate students with little technology background, graduate students with a general computer science technology background, and seasoned engineers who are experts in a particular field.

Figure 1 illustrates the proposed modular and integrated teaching approach: the class is divided into nine modules and integration experiments. Undergraduate and graduate students can both start with a general overview of autonomous driving technologies, but the undergraduate students may need more time to understand the basics of the technologies involved. They can then move on to localization, followed by traditional perception and perception with deep learning. Next, they can learn about the decision-making pipeline, including planning and control, motion planning, as well as end-to-end planning. Once the students are done with these, they can delve into client systems and cloud platforms. Finally, they can perform integration experiments to understand the interactions between these modules.

On the other hand, seasoned engineers with embedded system backgrounds can start with the general overview and then directly move on to the client systems module and get familiar with the new materials from the perspective of their comfort zone, thus allowing them to maintain a high interest level. They can then move on to the cloud platform, which also focuses on system design, and still stay within their comfort zone. Once they master these modules, they are equipped with enough background knowledge and confidence to learn the rest of the modules.

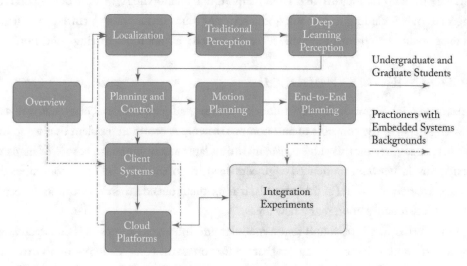

Figure 1: Example of our modular and integrated teaching approach.

3.2   MODULAR TEACHING MATERIALS

First, to cover all the major modules in the autonomous driving technology stack, we developed the current book, *Creating Autonomous Vehicle Systems*, now in its second edition. This is one of the first technical overviews of autonomous vehicles where we share our practical experiences creating autonomous vehicle systems. This book consists of nine chapters that provide an overview of autonomous vehicle systems followed by descriptions of localization technologies, the traditional techniques used for perception, deep learning-based techniques for perception, the planning and control sub-system (especially prediction and routing technologies) motion planning and feedback control for the planning and control subsystem, the reinforcement learning-based planning and control, the details of client systems design, and the details of cloud platforms for autonomous driving.

This book is aimed at students, researchers, and practitioners alike. For undergraduate or graduate students interested in autonomous driving, this book provides a comprehensive overview of the whole autonomous vehicle technology stack. For autonomous driving practitioners, it presents many practical techniques in implementing autonomous driving systems. It provides researchers with many references for an effective, deeper exploration of the various technologies.

Along with this book, in cooperation with IEEE Computer Society and O'Reilly, we have developed a series of online lectures to introduce each module [5, 6]. Along with the multimedia presentations, this allows students to easily acquire an in-depth understanding of a specific technology.

3.3   INTEGRATION EXPERIMENTAL PLATFORMS

A common problem of autonomous driving education is the lack of experimental platforms. In most autonomous driving classes, people use simulators to verify the performance of newly developed algorithms [11]. Nonetheless, the simulation approach fails to provide an environment for students to understand the interaction between different modules. On the other hand, using an autonomous driving vehicle as an experimental platform is not practical due to the cost, as a demonstration autonomous vehicle can easily cost over $800,000.

A straightforward way to perform integration experiments is to use mobile platforms, such as mobile phones. Nowadays, mobile phones usually consist of many sensors (GPS, IMU, cameras, etc.) as well as powerful heterogeneous computing platforms (with CPU, GPU, and DSP). Therefore, mobile phones can be used as a great integration experimental platform for localization and perception tasks. For instance, as demonstrated in the video "Perceptin Robot System Running on a Cell Phone [8], we have successfully implemented real-time localization, obstacle detection and avoidance, as well as planning and control functions on a Samsung Galaxy 7 mobile phone to drive a mobile robot at 5 miles per hour.

4.   CASE STUDIES

Several institutes and universities already have adopted or are in the process of adopting the teaching methodology as well as the experimental platforms described in this chapter. In this section, we present three case studies of applying the aforementioned teaching methodology and materials: the first one is an introduction to an autonomous driving class for undergraduate students with limited technology backgrounds; the second is a graduate-level embedded system class, in which we added a session on autonomous driving; and the last is a two-week professional training for seasoned engineers. These three case studies were carefully selected to demonstrate the flexibility of using the proposed approach to teach autonomous driving.

4.1   INTRODUCTION TO AUTONOMOUS DRIVING CLASS

The introduction to autonomous driving class for undergraduate and graduate students we have developed consists of 15–20 lectures depending on the length of the quarter or semester. Also, a 20-hour experiment session is required for integration experiments. Its overall purpose is to provide a technical overview on autonomous driving for students with basic experience in programming, algorithms, and operating systems.

Due to their limited background, we do not expect students to fully master all the modules, but we intend to maintain their interest level and equip them with the basic knowledge to delve into the modules in which they are particularly interested. To achieve this, we follow the approach shown in Figure 1 and divide the class into nine modules. To maintain a high level of interest, at the beginning of each session, we play a short video, such as "Creating Autonomous Vehicle Systems" [5], to provide a summary and demo of the technologies discussed. Then we move on to the details regarding of implementation of each technology. Also, throughout the class, we have the students use mobile phones to perform integration experiments. Specifically, for localization experiments, we first have the students extract real-time GPS localization data; then we have them improve their localization data by fusing IMU data with GPS data. For perception, we have the students install a deep learning framework, such as MXNET [12], onto their mobile phones and run simple object detection networks.

In addition, we developed multiple integration experiments for the students to gain a deep understanding of the interaction between different modules. Since it is an introductory class, for the advanced integration experiments, such as fusing GPS, IMU, and camera data to provide accurate location updates in real time, we did not provide enough technical background in the lectures. Thus, in order to accomplish these tasks, students not only need to perform their own research to get enough technical background but also spend significant time and effort to perform the experiments. Indeed, we did not expect any student to be able to accomplish the advanced tasks. To our surprise, 8% of the students were able to successfully accomplish these advanced tasks. These observations demonstrate that with a modular and integrated teaching approach, students not only obtain a comprehensive overview of the technologies but are also able to delve into the modules of their interests and become experts.

4.2   ADDING AUTONOMOUS DRIVING MATERIALS TO EMBEDDED SYSTEMS CLASS

We have also added a session in an existing graduate-level embedded systems class to explore how embedded systems technologies can be integrated in autonomous driving systems. The embedded systems class runs over a 20-week semester with 60 hours of lectures and 20 hours of experiments, in which we allocated 6 hours of lectures and 10 hours of experiments for autonomous driving contents. Interestingly, before we started this class, we checked with the students on whether they would start their engineering career in autonomous driving. Most students were highly interested but at the same time feared that autonomous driving was too complicated for them.

With the hope of easing the students' apprehension toward the design of autonomous driving systems, we placed the autonomous driving session at the end of the class, after the students grasped the basic skills of designing and implementing embedded systems with different software and hardware optimization techniques, such as heterogeneous computing. Prior to starting this case study session, all students' understanding of autonomous driving was limited to the conceptual level. Out of the 56 students enrolled, only 10 students were able to list some technologies involved in autonomous driving, such as localization and perception, but none understood the details of these technologies.

Due to the limited time available in the session, we first presented an overview on autonomous driving and then focused on two modules: localization and perception. We delved into two simple algorithm implementations, ORB-SLAM [9] for localization and SqueezeNet [10] for object detection. Next, we placed the students into groups of four to perform integration experiments with these algorithms on their Android cell phones and had them compare the performance of using CPU only vs. the performance of using heterogeneous computing components such as GPU and DSP. After completing the project, the students were asked to summarize their design choices and present their results in class. The presentations helped them understand the techniques used by other groups and they could learn from each other through the presentations.

The results were encouraging. First, it was very interesting to see different optimization strategies from different groups. Some groups prioritized computing resources for localization tasks to guarantee frequent position updates, whereas other groups prioritized computing resources for perception to guarantee real-time obstacle avoidance. Second, through this session, the apprehension toward autonomous driving went away, and the after-class survey indicated that 85% of the students would like to continue learning autonomous driving.

4.3   PROFESSIONAL TRAINING

For autonomous driving companies, one of the biggest challenges is the difficulty of recruiting autonomous driving engineers since there is only a very limited talent pool with autonomous driving experience. Therefore, it is crucial to develop a professional training session to quickly equip seasoned engineers with the technical knowledge to delve into one module of autonomous driving.

We worked closely with an autonomous driving company to quickly bring their engineers, most of whom have embedded systems and general software engineering backgrounds, up to speed. The challenges were three-fold: first, the training session was only two weeks long, not enough time to delve into the technology details; second, in this time span, we needed to place the engineers into different engineering roles, although they came from similar technology background; and third, confidence was a big issue for these engineers as they were concerned as to whether they could get a handle on the complexity of autonomous driving in a short amount of time.

To address these challenges, following the methodology presented in Figure 1, we had the engineers all start with the technology overview, followed by the client systems and the cloud platforms modules during the first week. Since the engineers came from embedded systems and general software engineering backgrounds, they were quite comfortable starting with these modules. Through system modules, they learned about the characteristics of different workloads as well as how to integrate them on embedded and cloud systems. In the second week, based on the engineers' performance in the first week as well as their interest levels toward different technologies, we assigned them to dig deeper into a specific module, such as perception, localization, or decision making.

For integration experiments, unlike in the undergraduate or graduate classes, the engineers were given the chance to work on an actual product after two weeks of training. In this training session, seven engineers were successfully added to the team; one was assigned to the sensing team, two were assigned to the perception team, two were added to the localization team, and two were assigned to the decision-making team. A demo of the development process is shown in the video "Perceptin Autonomous Vehicle Development" [7].

5. CONCLUSION

We are often asked what the most important technology in autonomous driving is. Our answer is always integration. As mentioned above and stressed here again, autonomous driving is not one single technology but rather a complex system integrating many technologies. However, before integration happens, one has to understand each technology module involved. Existing autonomous driving classes often focus on one or two technology modules, or directly have the students build a working autonomous vehicle, thus creating high entry barriers for students. Surprisingly, most students are indeed highly interested in autonomous driving, but it is the fear that they cannot handle the complexities involved that often drives them away.

To address this problem, we developed this modular and integrated approach to teaching autonomous driving. This approach breaks the complex autonomous driving system into different technology modules and first has the students understand each module. After the students grasp the basic concepts in each technology module, they are asked to perform integration experiments to help them understand the interactions between these modules.

We have successfully applied this methodology to three pilot case studies: an undergraduate-level introduction to autonomous driving class, a graduate-level embedded systems class with a session on autonomous driving, as well as a professional training session at an autonomous driving company. Although the students in these three pilot case studies have very diverse backgrounds, the modular teaching approach allowed us to adjust the order of modules flexibly to fit the needs of different students, and the integration experiments enabled the students to understand the interactions between different modules. This gave the students a comprehensive understanding of the modules as well as their interactions. In addition, our experiences showed that the proposed approach allowed students to start with their comfortable modules and then move on to other modules, therefore enabling the students to maintain a high interest level and good performance.

REFERENCES

[1]   Liu, S., Peng, J., and Gaudiot, J-L. 2017. Computer, drive my car! *Computer*, 50(1), pp. 8–8.

[2]   Liu, S., Tang, J., Zhang, Z., and Gaudiot, J-L. 2017. Computer architectures for autonomous driving. *Computer*, 50(8), pp. 18–25.

[3]   Liu, S., Tang, J., Wang, C., Wang, Q., and Gaudiot, J-L. 2017. A unified cloud platform for autonomous driving. *Computer*, (12), pp. 42–49.

[4]   Liu, S., Li, L., Tang, J., Wu, S., and Gaudiot, J-L. 2017. Creating autonomous vehicle systems. *Synthesis Lectures on Computer Science*, 6(1), pp. i–186. xxi

[5]   IEEE Computer Society, Creating Autonomous Vehicle Systems, accessed 1 Feb 2018, https://www.youtube.com/watch?v=B8A6BiRkNUw&t=93s. xxi, xxii

[6]   OReilly, Enabling Computer-Vision-Based Autonomous Vehicles, accessed 1 Feb 2018, https://www.youtube.com/watch?v=89giovpaTUE&t=434s. xxi

[7]   PerceptIn Autonomous Vehicle Development, accessed 1 Feb 2018, https://www.youtube.com/watch?v=rzRC57IXtRY. xxiv

[8]   PerceptIn, PerceptIn Robot System Running on a Cell Phone, accessed 1 Feb 2018, https://www.youtube.com/watch?v=Mib8SXacKEE. xxi

[9]   ORB-SLAM, accessed 1 Feb 2018, http://webdiis.unizar.es/~raulmur/orbslam/. xxiii

[10]   SqueezeNet, accessed 1 Feb 2018, https://github.com/DeepScale/SqueezeNet. xxiii

[11]   Tang, J., Liu, S., Wang, C., and Liu, C. 2017. Distributed simulation platform for autonomous driving. *International Conference on Internet of Vehicles (IOV)* 2017: pp. 190–200. xxi

[12]   Apache MXNET, accessed 1 Feb 2018, https://mxnet.apache.org/. xxii

[13]   Costa, V., Rossetti, R., and Sousa, A. 2017. Simulator for teaching robotics, ROS and autonomous driving in a competitive mindset. *International Journal of Technology and Human Interaction*, 13(4), p. 14. xviii

[14]   Arnaldi, N., Barone, C., Fusco, F., Leofante, F., and Tacchella, A. 2016. Autonomous driving and undergraduates: An affordable setup for teaching robotics, *Proceedings of the 3rd Italian Workshop on Artificial Intelligence and Robotics*, pp.,5–9, Genova, Italy, November 28, 2016. xviii

[15]   Artificial General Intelligence, accessed 1 Feb 2018, https://agi.mit.edu/. xviii

[16]   Deep Learning for Self-Driving Cars, accessed 1 Feb 2018, https://selfdrivingcars.mit.edu/. xviii

[17]   Artificial Intelligence: Principles and Techniques, accessed 1 Feb 2018, http://web.stanford.edu/class/cs221/. xviii

[18]   Paull, L., Tani, J., Zuber, M.T., Rus, D., How, J., Leonard, J., and Censi, A. 2016. Duckietown: An open, inexpensive and flexible platform for autonomy education and research, *IEEE International Conference on Robotics and Automation*, Singapore, May. 2017, pp. 1–8. DOI: 10.1007/978-3-319-55553-9_8. xviii

[19]   Karaman, S., Anders, A., Boulet, M., Connor, M.T., Abbott, J., Gregson, K.L., Guerra, W. J., Guldner, O.R., Mohamoud, M.M., Plancher, B.K., Robert, T-I., and Vivilecchia, J.R. 2017. Project-based, collaborative, algorithmic robotics for high school students: Programming self-driving race cars at MIT. *IEEE Integrated STEM Education Conference*, pp. 195–203, 2017. xviii

[20]   Tan, S. and Shen, Z., Hybrid problem-based learning in digital image processing: A case study, *IEEE Transactions on Education*, 2017, (99): pp. 1–9. xviii

[21]   Robotica 2017, accessed 1 Feb 2018, http://robotica2017.isr.uc.pt/index.php/en/competitions/major/autonomous-driving. xviii

[22]   Autonomous Driving Challenge, accessed 1 Feb 2018, http://www.autodrivechallenge.org/. xviii

[23]   NXP CUP Intelligent Car Racing, accessed 1 Feb 2018, https://community.nxp.com/groups/tfc-emea. xviii

CHAPTER 1

Introduction to Autonomous Driving

We are at the dawn of the future of autonomous driving. To understand what the future may be, we usually consult history, so let us start with that.

The beginning of information technology truly started in the 1960s, when Fairchild Semiconductors and Intel laid down the foundation layer of information technology by producing microprocessors, and as a side product, created Silicon Valley. Although microprocessor technologies greatly improved our productivity, the general public had limited access to these technologies. Then in the 1980s, Microsoft and Apple laid down the second layer of information technology by introducing Graphics User Interface, and realized the vision of "a PC/Mac in every home." Once everyone had access to computing power, in the 2000s, internet companies, represented by Google, laid down the third layer of information technology by connecting people and information. Through Google, for instance, the information providers and the information consumers can be indirectly connected. Then in the 2010s, the social network companies, such as Facebook and LinkedIn, laid down the fourth layer of information technology by effectively moving the human society to internet, and allowed people to directly connect to one another. After the population of the internet-based human society reached a significant scale, around 2015, the emergence of Uber and Airbnb laid down the fifth layer of information technology by providing services upon the internet-based human society, and forming an internet-based commerce society. Although Uber and Airbnb provided the means for us to efficiently access service providers through the internet, the services are still provided by humans.

1.1 AUTONOMOUS DRIVING TECHNOLOGIES OVERVIEW

As shown in Figure 1.1, autonomous driving is not one single technology, but rather a highly complex system that consists of many sub-systems. Let us break it into three major components: algorithms, including sensing, perception, and decision (which requires reasoning for complex cases); client systems, including the operating system and the hardware platform; and the cloud platform, including high-definition (HD) map, deep learning model training, simulation, and data storage.

The algorithm subsystem extracts meaningful information from sensor raw data to understand its environment and to make decisions about its future actions. The client systems integrate these algorithms together to meet real-time and reliability requirements. For example, if the camera generates data at 60 Hz, the client systems need to make sure that the longest stage of the processing pipeline takes less than 16 ms to complete. The cloud platform provides offline computing and storage capabilities for autonomous cars. With the cloud platform, we are able to test new algorithms, update HD map, and train better recognition, tracking, and decision models.

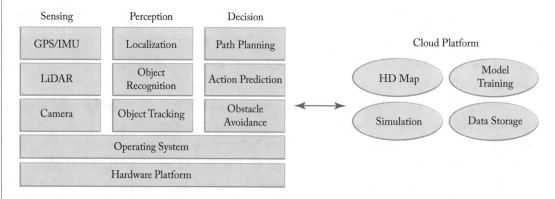

Figure 1.1: Autonomous driving system architecture overview.

1.2   AUTONOMOUS DRIVING ALGORITHMS

The algorithms component consists of sensing, that is extracting meaningful information from sensor raw data; perception, which is to localize the vehicle and to understand the current environment; and decision, in other words taking actions so as to reliably and safely reach target destinations.

1.2.1   SENSING

Normally, an autonomous car consists of several major sensors. Indeed, since each type of sensor presents advantages and drawbacks, in autonomous vehicles, the data from multiple sensors must be combined for increased reliability and safety. They can include the following.

- **GPS/IMU:** The GPS/IMU system helps the autonomous vehicle localize itself by reporting both inertial updates and a global position estimate at a high rate, e.g., 200 Hz. GPS is a fairly accurate localization sensor, but its update rate is slow, at about only 10 Hz, and thus not capable of providing real-time updates. However, IMU errors accumulate over time, leading to a corresponding degradation in the position estimates. Nonetheless, an IMU can provide updates more frequently, at or higher than 200 Hz. This should satisfy the real-time requirement. By combining both GPS and IMU, we can provide accurate and real-time updates for vehicle localization.

- **LiDAR:** LiDAR is used for mapping, localization, and obstacle avoidance. It works by bouncing a beam off surfaces and measures the reflection time to determine distance. Due to its high accuracy, LiDAR can be used to produce HD maps, to localize a moving vehicle against HD maps, to detect obstacle ahead, etc. Normally, a LiDAR unit, such as Velodyne 64-beam laser, rotates at 10 Hz and takes about 1.3 million readings per second.

- **Cameras:** Cameras are mostly used for object recognition and object tracking tasks such as lane detection, traffic light detection, and pedestrian detection, etc. To enhance autonomous vehicle safety, existing implementations usually mount eight or more 1080p cameras around the car, such that we can use cameras to detect, recognize, and track objects in front of, behind, and on both sides of the vehicle. These cameras usually run at 60 Hz, and, when combined, would generate around 1.8 GB of raw data per second.

- **Radar and Sonar:** The radar and sonar system is mostly used for the last line of defense in obstacle avoidance. The data generated by radar and sonar shows the distance as well as velocity from the nearest object in front of the vehicle's path. Once we detect that an object is not far ahead, there may be a danger of a collision, then the autonomous vehicle should apply the brakes or turn to avoid the obstacle. Therefore, the data generated by radar and sonar does not require much processing and usually is fed directly to the control processor, and thus not through the main computation pipeline, to implement such "urgent" functions as swerving, applying the brakes, or pre-tensioning the seatbelts.

1.2.2   PERCEPTION

The sensor data is then fed into the perception stage to provide an understanding of the vehicle's environment. The three main tasks in autonomous driving perception are localization, object detection, and object tracking.

GPS/IMU can be used for localization, and, as mentioned above, GPS provides fairly accurate localization results but with a comparatively low update rate, while an IMU provides very fast updates at a cost of less accurate results. We can thus use Kalman Filter techniques to combine the advantages of the two and provide accurate and real-time position updates. As shown in Figure 1.2, it works as follows: the IMU updates the vehicle's position every 5 ms, but the error accumulates with time. Fortunately, every 100 ms, a GPS update is received, which helps correct the IMU error. By running this propagation and update model, the GPS/IMU combination can generate fast and accurate localization results. Nonetheless, we cannot solely rely on this combination for localization for three reasons: (1) the accuracy is only about one meter; (2) the GPS signal has multipath problems, meaning that the signal may bounce off buildings, introducing more noise; and (3) GPS requires an unobstructed view of the sky and would thus not work in environments such as tunnels.

Localization

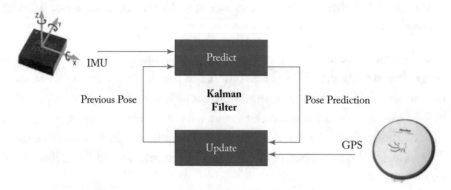

Figure 1.2: GPS/IMU localization.

Cameras can be used for localization too. Vision-based localization can be implemented as the following simplified pipeline: (1) by triangulating stereo image pairs, we first obtain a disparity map which can be used to derive depth information for each point; (2) by matching salient features between successive stereo image frames, we can establish correlations between feature points in different frames. We could then estimate the motion between the past two frames; and also, (3) by comparing the salient features against those in the known map, we could also derive the current position of the vehicle. However, such a vision-based localization approach is very sensitive to lighting conditions and, thus, this approach alone would not be reliable.

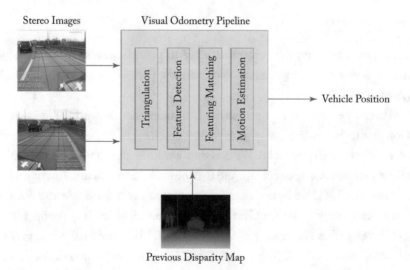

Figure 1.3: Stereo visual odometry.

This is why LiDAR approaches typically have recourse to particle filter techniques. The point clouds generated by LiDAR provide a "shape description" of the environment, but it is hard to differentiate individual points. By using a particle filter, the system compares a specific observed shape against the known map to reduce uncertainty. To localize a moving vehicle relative to these maps, we could apply a particle filter method to correlate the LIDAR measurements with the map. The particle filter method has been demonstrated to achieve real-time localization with 10-cm accuracy and to be effective in urban environments. However, LiDAR has its own problem: when there are many suspended particles in the air, such as rain drops and dust, the measurements may be extremely noisy. Therefore, as shown in Figure 1.4, to achieve reliable and accurate localization, we need a sensor-fusion process to combine the advantages of all sensors.

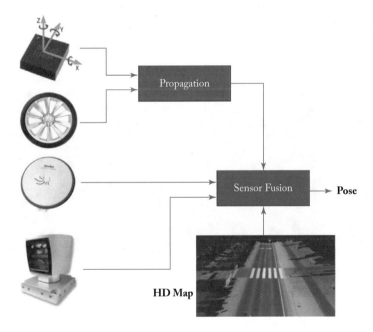

Figure 1.4: Sensor-fusion localization pipeline.

1.2.3   OBJECT RECOGNITION AND TRACKING

Originally LiDAR was used mostly to perform object detection and tracking tasks in Autonomous Vehicles, since LiDAR provides very accurate depth information. In recent years, however, we have seen the rapid development of Deep Learning technology, which achieves significant object detection and tracking accuracy. Convolution Neural Network (CNN) is a type of Deep Neural Network (DNN) that is widely used in object recognition tasks. A general CNN evaluation pipeline usually consists of the following layers. (1) The Convolution Layer contains different filters to extract different features from the input image. Each filter contains a set of "learnable" parameters that will be derived after the training stage. (2) The Activation Layer decides whether to activate the target neuron or not. (3) The Pooling Layer reduces the spatial size of the representation to reduce the number of parameters and consequently the computation in the network. (4) The Fully Connected Layer where neurons have full connections to all activations in the previous layer.

Object tracking refers to the automatic estimation of the trajectory of an object as it moves. After the object to track is identified using object recognition techniques, the goal of object tracking is to automatically track the trajectory of the object subsequently. This technology can be used to track nearby moving vehicles as well as people crossing the road to ensure that the current vehicle does not collide with these moving objects. In recent years, deep learning techniques have demonstrated advantages in object tracking compared to conventional computer vision techniques. Specifically, by using auxiliary natural images, a stacked Auto-Encoder can be trained offline to learn generic image features that are more robust against variations in viewpoints and vehicle positions. Then, the offline trained model can be applied for online tracking.

Figure 1.5: Object recognition and tracking [34], used with permission.

1.2.4 ACTION

Based on the understanding of the vehicle's environment, the decision stage can generate a safe and efficient action plan in real time.

Action Prediction

One of the main challenges for human drivers when navigating through traffic is to cope with the possible actions of other drivers which directly influence their own driving strategy. This is especially true when there are multiple lanes on the road or when the vehicle is at a traffic change point. To make sure that the vehicle travels safely in these environments, the decision unit generates predictions of nearby vehicles, and decides on an action plan based on these predictions. To predict actions of other vehicles, one can generate a stochastic model of the reachable position sets of the other traffic participants, and associate these reachable sets with probability distributions.

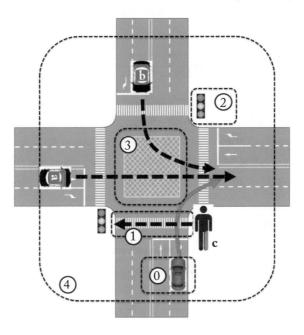

Figure 1.6: Action prediction.

Path Planning

Planning the path of an autonomous, agile vehicle in a dynamic environment is a very complex problem, especially when the vehicle is required to use its full maneuvering capabilities. A brute force approach would be to search all possible paths and utilize a cost function to identify the best path. However, the brute force approach would require enormous computation resources and may be unable to deliver navigation plans in real-time. In order to circumvent the computational complexity of deterministic, complete algorithms, probabilistic planners have been utilized to provide effective real-time path planning.

Obstacle Avoidance

As safety is the paramount concern in autonomous driving, we usually employ at least two levels of obstacle avoidance mechanisms to ensure that the vehicle would not collide with obstacles. The first level is proactive, and is based on traffic predictions. At runtime, the traffic prediction mechanism generates measures like time to collision or predicted minimum distance, and based on this information, the obstacle avoidance mechanism is triggered to perform local path re-planning. If the proactive mechanism fails, the second level, the reactive mechanism, using radar data, would take over. Once radar detects an obstacle ahead of the path, it would override the current control to avoid the obstacles.

1.3   AUTONOMOUS DRIVING CLIENT SYSTEM

The client systems integrate the above-mentioned algorithms together to meet real-time and reliability requirements. Some of the challenges are as follows: the system needs to ensure that the processing pipeline is fast enough to process the enormous amount of sensor data generated; if a part of the system fails, it must be sufficiently robust to recover from the failure; and, in addition, it needs to perform all the computation under strict energy and resource constraints.

1.3.1   ROBOT OPERATING SYSTEM (ROS)

ROS is a powerful distributed computing framework tailored for robotics applications, and it has been widely used. As shown in Figure 1.7, each robotic task (such as localization), is hosted in a ROS node. ROS nodes can communicate with each other through topics and services. It is a great operating system for autonomous driving except that it suffers from several problems: (1) reliability: ROS has a single master and no monitor to recover failed nodes; (2) performance: when sending out broadcast messages, it duplicates the message multiple times, leading to performance degradation; and (3) security: it has no authentication and encryption mechanisms. Although ROS 2.0 promised to fix these problems, ROS 2.0 itself has not been extensively tested and many features are not yet available. Therefore, in order to use ROS in autonomous driving, we need to solve these problems first.

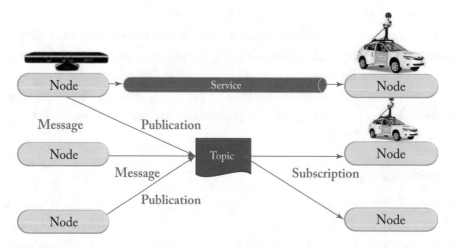

Figure 1.7: Robot operating system (ROS).

Reliability

The current ROS implementation has only one master node, when the master node crashes, the whole system would crash. This does not meet the safety requirement for autonomous driving. To fix this problem, we implemented a ZooKeeper-like mechanism in ROS. As shown in Figure 1.8, in this design, we have a main master node and a backup master node. In the case of main node failure, the backup node would take over, making sure that the system still runs without hiccups. In addition, this ZooKeeper mechanism monitors and restarts any failed nodes, making sure the whole ROS system is reliable.

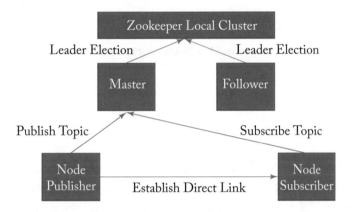

Figure 1.8: Zookeeper for ROS.

Performance

Performance is another problem with the current ROS implementation. The ROS nodes communicate often, and it is imperative to ensure the communication between nodes is efficient. First, it goes through the loop-back mechanism when local nodes communicate with each other. Each time it goes through the loop-back pipeline, a 20 ms overhead is introduced. To eliminate this overhead, for local node communication overhead, we used shared memory mechanism such that the message does not have to go through the TCPIP stack to get to the destination node. Second, when a ROS node broadcasts a message, the message gets copied multiple times, consuming significant bandwidth in the system. As shown in Figure 1.9, by switching to multicast mechanism, we greatly improved the throughput of the system.

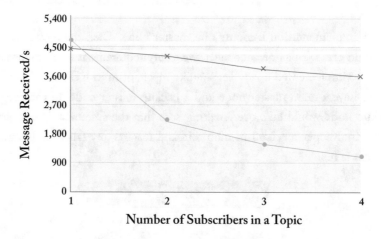

Figure 1.9: Multicast vs. broadcast in ROS.

Security

As we know, security is the most critical concern for ROS. Now imagine two scenarios: in the first scenario, a ROS node is kidnapped and continuously allocates memory until the system runs out of memory and starts killing other ROS nodes. In this scenario, the hacker successfully crashes the system. In the second scenario, since by default ROS messages are not encrypted, a hacker can easily eavesdrop the message between nodes and applies man-in-the-middle attacks. To fix the first problem, we can use Linux Container (LXC) to restrict the amount of resources used by each node, and also to provide a sandbox mechanism to protect the node from each other, therefore effectively preventing resource leaking. To fix the second problem, we can encrypt messages in communication, preventing messages being eavesdropped.

1.3.2   HARDWARE PLATFORM

To understand the challenges in designing hardware platform for autonomous driving, let us examine the computing platform implementation from a leading autonomous driving company. It consists of two compute boxes, each equipped with an Intel Xeon E5 processor and four to eight Nvidia K80 GPU accelerators. The second compute box performs exactly the same tasks and is used for reliability: in case the first box fails, the second box can immediately take over. In the worst case, when both boxes run at their peak, this would mean over 5000 W of power consumption which would consequently generate enormous amount of heat. Also, each box costs $20,000–$30,000, making the whole solution unaffordable to average consumers.

The power, heat dissipation, and cost requirements of this design prevents autonomous driving to reach the general public. To explore the edges of the envelope and understand how well an autonomous driving system could perform on an ARM mobile SoC, we implemented a simplified, vision-based autonomous driving system on a ARM-based mobile SoC with peak power consumption of 15 W. As it turns out, the performance was close to our requirements: the localization pipeline was capable of processing 25 images per second, almost keeping up with an image generation rate of 30 images per second. The deep learning pipeline was capable of performing 2–3 object recognition tasks per second. The planning and control pipeline could plan a path within 6 ms. With this system, we were able to drive the vehicle at around 5 mph without any loss of localization.

1.4   AUTONOMOUS DRIVING CLOUD PLATFORM

Autonomous vehicles are mobile systems, and therefore they need a cloud platform to provides supports. The two main functions provided by the cloud include distributed computing, and distributed storage. It has several applications, including simulation, which is used to verify new algorithms; HD map production; and deep learning model training. To build such a platform, we used Spark for distributed computing, OpenCL for heterogeneous computing, and Alluxio for in-memory storage. We have managed to deliver a reliable, low-latency, and high-throughput autonomous driving cloud by integrating Spark, OpenCL, and Alluxio together.

1.4.1   SIMULATION

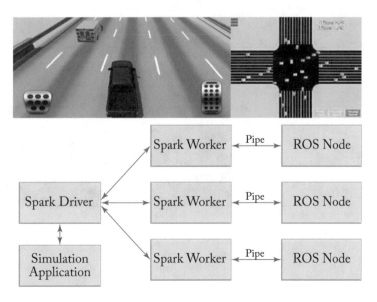

Figure 1.10: Spark and ROS-based simulation platform.

The first application of such a system is simulation. When we develop a new algorithm, and we need to test it thoroughly before we can deploy it on cars. If we were to test it on real cars, the cost would be enormous and the turn-around time would be too long. Therefore, we usually test it on simulators, such as replaying data through ROS nodes. However, if we were to test the new algorithm on a single machine, either it is going to take too long, or we do not have enough test coverage. As shown in Figure 1.10, to solve this problem, we have developed a distributed simulation platform. Such that we use Spark to manage distributed computing nodes, and on each node, we run a ROS replay instance. In an autonomous driving object recognition test set that we used, it took 3 h to run on a single server; by using the distributed system we developed, the test finished within 25 min when we scaled to 8 machines.

1.4.2   HD MAP PRODUCTION

As shown in Figure 1.11, HD map production is a complex process that involves many stages, including raw data processing, point cloud production, point cloud alignment, 2D reflectance map generation, HD map labeling, as well as the final map generation. Using Spark, we connected all these stages together in one Spark job. More importantly, Spark provides an in-memory computing mechanism, such that we do not have to store the intermediate data in hard disk, thus greatly reducing the performance of the map production process.

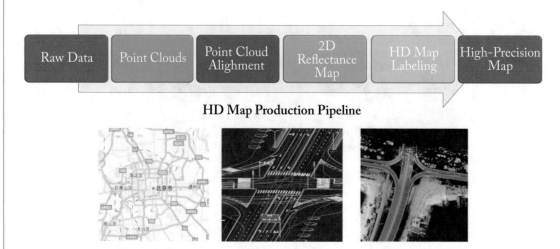

Figure 1.11: Cloud-based HD Map production.

1.4.3   DEEP LEARNING MODEL TRAINING

As we use different deep learning models in autonomous driving, it is imperative to provide updates to continuously improve the effectiveness and efficiency of these models. However, since the amount of raw data generated is enormous, we would not be able to achieve fast model training using single servers. To approach this problem, we have developed a highly scalable distributed deep learning system using Spark and Paddle (a deep learning platform recently open-sourced by Baidu). As shown in Figure 1.12, in the Spark driver we manage a Spark context and a Paddle context, and in each node, the Spark executor hosts a Paddler trainer instance. On top of that, we use Alluxio as a parameter server for this system. Using this system, we have achieved linear performance scaling as we added more resources, proving that the system is highly scalable.

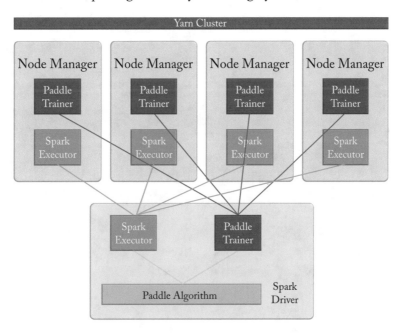

Figure 1.12: Distributed deep learning model training system.

1.5   IT IS JUST THE BEGINNING

Again, autonomous driving, or artificial intelligence in general, is not one single technology, it is an integration of many technologies. It demands innovations in algorithms, system integrations, cloud platforms. It is just the beginning and there are tons of opportunities in this era, I foresee that in 2020, we will officially start this AI-era and start seeing more and more AI-based products in the market.

CHAPTER   2

Autonomous Vehicle Localization

Abstract

>*For autonomous vehicles, one of the most critical tasks is localization, i.e., the accurate and real-time determination of the unit's position. In this chapter we first study different localization techniques, including GNSS, LiDAR and High-Definition Maps, Visual Odometry, and other Dead Reckoning sensors. We also look into several real-world examples of applying sensor fusion techniques to combine multiple sensors to provide more accurate localization.*

2.1   LOCALIZATION WITH GNSS

When human drive cars, we usually rely on the global navigation satellite system (GNSS) for localization. When it comes to autonomous vehicle localization, we also start with GNSS. In this section, we delve into the details of GNSS technologies and understand the pros and cons of GNSS when applying to autonomous driving.

2.1.1   GNSS OVERVIEW

The GNSS consist of several satellite systems: GPS, GLONASS, Galileo, and BeiDou. Here we use GPS as an example to provide an overview of GNSS. GPS provides coded satellite signals that can be processed in a GPS receiver, allowing the receiver to estimate position, velocity and time [1]. For this to work, GPS requires four satellite signals to compute positions in three dimensions and the time offset in the receiver clock. The deployment of these GPS satellites are dispersed in six orbital planes on almost circular orbits with an altitude of about 20,200 km above the surface of the Earth, inclined by 55° with respect to the equator and with orbital periods of approximately 11 hr 58 min.

The generated signals on board the satellites are derived from generation of a fundamental frequency fo=10.23 MHz [1]. The signal is time stamped with atomic clocks with inaccuracy in the range of only 10–13 s over a day. Two carrier signals in the L-band, denoted L1 and L2, are generated by integer multiplications of fo. The carriers L1 and L2 are bi-phase modulated by codes to provide satellite clock readings to the receiver and transmit information such as the orbital parameters. The codes consist of a sequence with the states +1 or −1, corresponding to the binary values 0 or 1. The bi-phase modulation is performed by a 180° shift in the carrier phase whenever a change in the code state occurs. The satellite signals contain information on the satellite orbits, orbit perturbations, GPS time, satellite clock, ionospheric parameters, and system status messages, etc. The navigation message consists of 25 frames with each frame containing 1,500 bit and each frame is subdivided into 5 sub-frames with 300 bit.

The next critical piece of the GNSS system is the definition of reference coordinate system, which is crucial for the description of satellite motion, the modeling of observable and the interpretation of results. For GNSS to work, two reference systems are required: (a) space-fixed, inertial reference system for the description of satellite motion; and (b) earth-fixed, terrestrial reference system for the positions of the observation stations and for the description of results from satellite geodesy. The two systems are used and the transformation parameters between the space fixed and earth fixed are well known and used directly in the GNSS receiver and post processing software to compute the position of the receivers in the earth fixed system. Terrestrial reference system is defined by convention with three axes, where Z-axis coincides with the earth rotation axis as defined by the Conventional International Origin. The X-axis is associated with the mean Greenwich meridian, and the Y-axis is orthogonal to both Z and X axes and it completes the right-handed coordinate system. GPS has used the WGS84 as a reference system and with WGS84 associated a geocentric equipotential ellipsoid of revolution [2].

In recent years, the emergence of GNSS receivers supporting multiple constellations has kept steady pace with the increasing number of GNSS satellites in the sky in the past decade. With advancements in newer GNSS constellations, almost 100% of all new devices are expected to support multiple constellations. The benefits of supporting multiple constellations include increased availability, particularly in areas with shadowing; increased accuracy, more satellites in view improves accuracy; and improved robustness, as independent systems are harder to spoof.

2.1.2   GNSS ERROR ANALYSIS

Ideally, with GNSS, we can get perfect localization results with no error at all. However, there are multiple places where error can be introduced in GNSS. In this subsection, we review these potential error contributors.

- **Satellite Clocks:** Any tiny amount of inaccuracy of the atomic clocks in the GNSS satellites can result in a significant error in the position calculated by the receiver. Roughly, 10 ns of clock error results in 3 m of position error.

- **Orbit Errors:** GNSS satellites travel in very precise, well-known orbits. However, like the satellite clock, the orbits do vary a small amount. When the satellite orbit changes, the ground control system sends a correction to the satellites and the satellite ephemeris is updated. Even with the corrections from the GNSS ground control system, there are still small errors in the orbit that can result in up to ±2.5 m of position error.

- **Ionospheric Delay:** The ionosphere is the layer of atmosphere between 80 km and 600 km above the earth. This layer contains electrically charged particles called ions. These ions delay the satellite signals and can cause a significant amount of satellite position error (typically ±5 m). Ionospheric delay varies with solar activity, time of year, season, time of day and location. This makes it very difficult to predict how much ionospheric delay is impacting the calculated position. Ionospheric delay also varies based on the radio frequency of the signal passing through the ionosphere.

- **Tropospheric Delay:** The troposphere is the layer of atmosphere closest to the surface of the Earth. Variations in tropospheric delay are caused by the changing humidity, temperature and atmospheric pressure in the troposphere. Since tropospheric conditions are very similar within a local area, the base station and rover receivers experience very similar tropospheric delay. This allows RTK GNSS to compensate for tropospheric delay, which will be discussed in the next subsection.

- **Multipath:** Multipath occurs when a GNSS signal is reflected off an object, such as the wall of a building, to the GNSS antenna. Because the reflected signal travels farther to reach the antenna, the reflected signal arrives at the receiver slightly delayed. This delayed signal can cause the receiver to calculate an incorrect position.

We have summarized the error ranges of these contributing sources in Figure 2.1. For a more detailed discussion of these errors, please refer to [3, 4, 5, 6].

Contributing Source	Error Range
Satellite Clocks	±2 m
Orbit Errors	±2.5 m
Inospheric Delays	±5 m
Tropospheric Delays	±0.5 m
Receiver Noise	±0.3 m
Multipath	±1 m

Figure 2.1: GNSS system errors (based on [3]).

2.1.3   SATELLITE-BASED AUGMENTATION SYSTEMS

Satellite-Based Augmentation Systems (SBAS) complement existing GNSS to reduce measurement errors. SBAS compensate for certain disadvantages of GNSS in terms of accuracy, integrity, continuity, and availability. The SBAS concept is based on GNSS measurements by accurately located reference stations deployed across an entire continent. The GNSS errors are then transferred to a computing center, which calculates differential corrections and integrity messages that are then broadcasted over the continent using geostationary satellites as an augmentation or overlay of the original GNSS message. SBAS messages are broadcast via geostationary satellites able to cover vast areas.

Several countries have implemented their own satellite-based augmentation system. Europe has the European Geostationary Navigation Overlay Service (EGNOS) which mainly covers the Europe. The U.S. has its Wide Area Augmentation System (WAAS). China has launched the BeiDou System (BDS) that provides its own SBAS implementation. Japan is covered by its Multi-functional Satellite Augmentation System (MSAS). India has launched its own SBAS program named GPS and GEO Augmented Navigation (GAGAN) to cover the Indian subcontinent. All of the systems comply with a common global standard and are therefore all compatible and interoperable.

Note that most commercial GNSS receivers provides SBAS function. In detail, the WAAS specification requires it to provide a position accuracy of 7.6 m or less for both lateral and vertical measurements, at least 95% of the time. Actual performance measurements of the system at specific locations have shown it typically provides better than 1.0 m laterally and 1.5 m vertically throughout most of the U.S.

2.1.4   REAL-TIME KINEMATIC AND DIFFERENTIAL GPS

Based on our experiences, most commercially available multi-constellation GNSS systems provide a localization accuracy no better than a 2-m radius. While this may be enough for human drivers, in order for an autonomous vehicle to follow a road, it needs to know where the road is. To stay in a specific lane, it needs to know where the lane is. For an autonomous vehicle to stay in a lane, the localization requirements are in the order of decimeters. Fortunately, Real-Time Kinematic (RTK) and Differential GNSS does provide decimeter level localization accuracy. In this subsection, we study how RTK and Differential GNSS works.

RTK GNSS achieves high accuracy by reducing errors in satellite clocks, imperfect orbits, inospheric delays, and trophospheric delays. Figure 2.2 shows the basic concept behind RTK GNSS, a good way to correct these GNSS errors is to set up a GNSS receiver on a station whose position is known exactly, a base station. The base station receiver calculates its position from satellite data and compares that position with its actual known position, and identifies the difference. The resulting error corrections can then be communicated from the base to the vehicle.

In detail, RTK uses carrier-based ranging and provides ranges (and therefore positions) that are orders of magnitude more precise than those available through code-based positioning. Code-based positioning is one processing technique that gathers data via a coarse acquisition code receiver, which uses the information contained in the satellite pseudo-random code to calculate positions. After differential correction, this processing technique results in 5-m accuracy. Carrier-based ranging is another processing technique that gathers data via a carrier phase receiver, which uses the radio carrier signal to calculate positions. The carrier signal, which has a much higher frequency than the pseudo-random code, is more accurate than using the pseudo-random code alone. The pseudo-random code narrows the reference then the carrier code narrows the reference even more. After differential correction, this processing technique results in sub-meter accuracy. Under carrier-based ranging, the range is calculated by determining the number of carrier cycles between the satellite and the vehicle, then multiplying this number by the carrier wavelength. The calculated ranges still include errors from such sources as satellite clock and ephemerides, and ionospheric and tropospheric delays. To eliminate these errors and to take advantage of the precision of carrier-based measurements, RTK performance requires measurements to be transmitted from the base station to the vehicle.

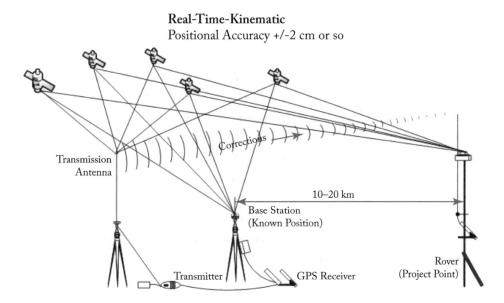

Figure 2.2: RTK GNSS (based on [46]).

With RTK GNSS, vehicles determine their position using algorithms that incorporate ambiguity resolution and differential correction. The position accuracy achievable by the vehicle depends on its distance from the base station and the accuracy of the differential corrections. Corrections are as accurate as the known location of the base station and the quality of the base station's satellite observations. Therefore, site selection is critical for minimizing environmental effects such as interference and multipath, as is the quality of the base station and vehicle receivers and antennas.

2.1.5   PRECISE POINT POSITIONING

Although RTK GNSS system provides sub-decimeter-level accuracy required to meet autonomous driving requirements, this solution often requires the users to deploy their own base stations, which are expensive to maintain. In this subsection, we study how Precise Point Positioning (PPP) GNSS system can help mitigate the problem [7, 8].

Figure 2.3 shows how a PPP GNSS solution works. Many reference stations are deployed worldwide, and these stations receive precise reference satellite orbits and reference GNSS satellite clocks in real time. These reference stations then calculate the corrections that should be applied to the satellite localization results. Once the corrections are calculated, they are delivered to the end users via satellite or over the Internet. The precise satellite positions and clocks minimize the satellite clock errors and orbit errors. We can then apply a dual-frequency GNSS receiver to remove the first-order effect of the ionosphere that is proportional to the carrier wave frequency. Therefore, the first-order ionospheric delay can be totally eliminated by using a combination of dual-frequency GNSS measurements. In addition, the tropospheric delay is corrected using the UNB model [9]: to achieve further accuracy, the residual tropospheric delay is estimated when estimating position and other unknowns [10]. By combining these techniques, PPP is capable of providing position solutions at the decimeter to centimeter level.

Specifically, the PPP algorithm uses as input code and phase observations from a dual-frequency receiver, and precise satellite orbits and clocks, in order to calculate precise receiver coordinates and clock. The observations coming from all the satellites are processed together in a filter, such as an Extended Kalman Filter (EKF). Position, receiver clock error, tropospheric delay, and carrier-phase ambiguities are estimated EKF states. EKF minimizes noise in the system and enables estimating position with centimeter-level accuracy. The estimates for the EKF states are improved with successive GNSS measurements, until they converge to stable and accurate values.

PPP differs from RTK positioning in the sense that it does not require access to observations from one or more close base stations and that PPP provides an absolute positioning instead of the location relative to the base station as RTK does. PPP just requires precise orbit and clock data, computed by a ground-processing center with measurements from reference stations from a relatively sparse station network. Note that PPP involves only a single GPS receiver and, therefore, no reference stations are needed in the vicinity of the user. Therefore, PPP can be regarded as a global position approach because its position solutions referred to a global reference frame. Hence, PPP provides much greater positioning consistency than the RTK approach in which position solutions are relative to the local base station or stations. Also, PPP is similar in structure to an SBAS system. Compared to SBAS, the key advantage provided by PPP is that it requires the availability of precise reference GNSS orbits and clocks in real-time, and thus achieving up to centimeter-level accuracy while SBAS only achieves meter-level accuracy. In addition, PPP systems allow a single correction stream to be used worldwide, while SBAS systems are regional.

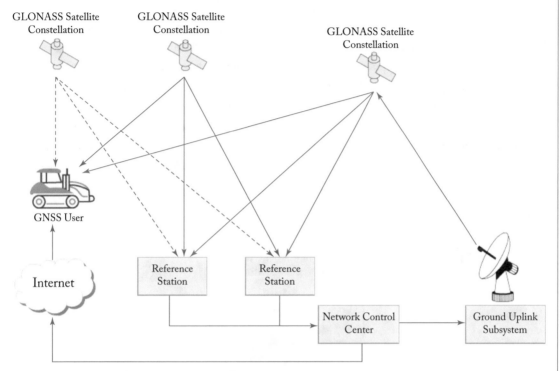

Figure 2.3: **PPP GNSS.** Courtesy of NovAtel, Inc., used with permission.

One major problem faced by the PPP solution is that, in order to resolve any local biases such as the atmospheric conditions, multipath environment and satellite geometry, a long period of time (30 min) is usually required in order to converge to decimeter accuracy. Currently, there exist several consolidated post-processing PPP services. Conversely, real-time PPP systems are in an incipient development phase [11, 12, 13].

2.1.6   GNSS INS INTEGRATION

In the previous subsections, we have studied different generations of GNSS technologies, in this subsection we explore how inertial data can be utilized to improve GNSS localization methods [14]. An Inertial Navigation System (INS) uses rotation and acceleration information from an Inertial Measurement Unit (IMU) to compute a relative position over time. A typical, six-axis IMU is made up of six complimentary sensors arrayed on three orthogonal axes. On each of the three axes is coupled an accelerometer and a gyroscope. The accelerometers measure linear acceleration and the gyroscopes measure rotational acceleration. With these sensors, an IMU can measure its precise relative movement in 3D space. The INS uses these measurements to calculate position and velocity. In addition, the IMU measurements provide angular velocities about the three axes, which can be used to deduce local attitudes (roll, pitch, and azimuth).

Typically, INS systems run at a rate of 1 KHz, providing very frequent position updates. However, INS systems also suffer from several drawbacks: First, an INS provides only a relative solution from an initial start point. This initial start point must be provided to the INS. Second, and more critically, navigating in 3D space with an IMU is effectively a summation (or integration) of hundreds/thousands of samples per second during which time the errors are also being accumulated. This means that an uncorrected INS system will drift from the true position quickly if without an external reference to correct it. Thus, when using INS to perform localization tasks, it is imperative to provide an accurate external reference to the INS, allowing it to minimize the localization errors using a mathematical filter, such as a Kalman Filter.

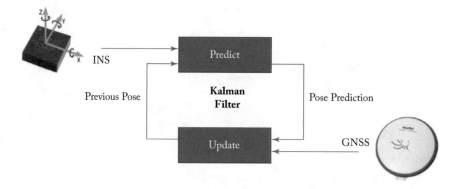

Figure 2.4: GNSS/INS fusion with Kalman Filter.

As shown in Figure 2.4, that external reference can be provided by GNSS. GNSS provides an absolute set of coordinates that can be used as the initial start point. Also, GNSS provides continuous positions and velocities thereafter to update the INS filter estimates. When GNSS signal is compromised due to signal obstructions (such as in the case of driving through a tunnel), the INS system can be used to localize the vehicle in a short period of time.

2.2   LOCALIZATION WITH LIDAR AND HIGH-DEFINITION MAPS

Most commercial autonomous vehicles prototypes, including Waymo, Baidu, BMW, etc., rely on LiDAR and HD Maps for localization. In this section, we study how LiDAR and HD Maps work, and how to combine the two to provide accurate localization for autonomous vehicles.

2.2.1   LIDAR OVERVIEW

In this subsection, we provide an overview of LiDAR technologies. LiDAR stands for Light Detection And Ranging, which measures distance to a target by illuminating that target with a pulsed laser light, and measuring the reflected pulses with a sensor [15]. Differences in laser return times and wavelengths can then be used to make digital 3D-representations of the target. The basic working principle behind LiDAR is as follows: first, a LiDAR instrument fires rapid pulses of laser light at a surface, some at up to 150,000 pulses/s. Then, a sensor on the instrument measures the amount of time it takes for each pulse to bounce back. As light moves at a constant and known speed so the LiDAR instrument can calculate the distance between itself and the target with high accuracy. By repeating this in quick succession the instrument builds up a complex "map" of the surface it is measuring.

Generally, there are two types of LiDAR detection methods: incoherent detection (also known as direct energy detection) and coherent detection [16]. Coherent systems are best for Doppler or phase sensitive measurements and generally use optical heterodyne detection, a method of extracting information encoded as modulation of the phase or frequency of electromagnetic radiation. This allows them to operate at much lower power but has the expense of more complex transceiver requirements. In detail, when incoherent light is emitted, it spreads in all directions. On the contrary, coherent light uses highly specialized diodes which generate energy at or near the optical portion of the electromagnetic spectrum, meaning that all the individual energy waves are moving in the same direction, resulting in much lower power consumption.

In both coherent and incoherent types of LiDAR, there exist two main pulse models: high-energy and micro-pulse systems. High energy systems emit high power lights and can be harmful to human eyes, these systems are commonly used for atmospheric research where they are often used for measuring a variety of atmospheric parameters such as the height, layering and density of clouds, cloud particles properties, temperature, pressure, wind, humidity, and trace gas concentration. On the contrary, micro-pulse systems are lower-powered and are classed as *eye-safe*, allowing them to be used with little safety precautions. In its original design [17], the micro-pulse LiDAR transmitter is a diode pumped micro-joule pulse energy, high-repetition-rate laser. Eye safety is obtained through beam expansion. The receiver uses a photon counting solid-state Geiger mode avalanche photodiode detector. The LiDAR devices used in autonomous driving are mostly coherent micro-pulse systems such that they meet category one laser safety requirements, which are the safest in all categories.

The lasers used in LiDAR can be categorized by their wavelength. 600–1000 nm lasers are most commonly used and usually its maximum power is limited to meet category one requirements. Lasers with a wavelength of 1,550 nm are also commonly used as they can be used for longer range and lower accuracy purposes. In addition, 1,550 nm wavelength laser does not show under night-vision goggles and is therefore well suited to military applications. Airborne LiDAR systems use 1064 nm diode pumped YAG lasers while Bathymetric systems use 532 nm double diode pumped YAG lasers which penetrate water with much less attenuation than the airborne 1,064 nm version. Better resolution can be achieved with shorter pulses provided the receiver detector and electronics have sufficient bandwidth to cope with the increased data flow.

A typical LiDAR system consists of two groups of major components, the laser scanners and the laser receivers. The speed at which images can be generated is affected by the speed at which it can be scanned into the system. A variety of scanning methods are available for different purposes such as azimuth and elevation, dual oscillating plane mirrors, dual axis scanner, and polygonal mirrors. They type of optic determines the resolution and range that can be detected by a system [18, 19]. Laser receivers read and record the signal being returned to the system. There are two main types of laser receiver technologies, silicon avalanche photodiodes, and photomultipliers [20].

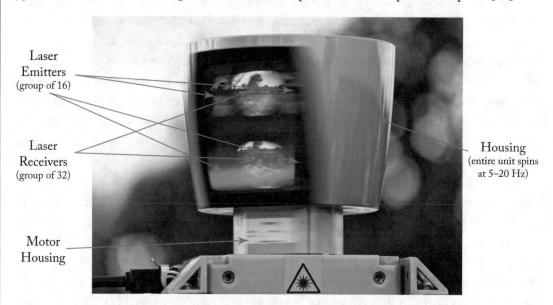

Figure 2.5: Velodyne HDL-64 LiDAR [45].

Figure 2.5 shows a Velodyne HDL-64 LiDAR, which is widely used in autonomous vehicles. It utilizes 64 LiDAR channels aligned from +2.0° to -24.9° for a vertical field of view of 26.9° and delivers a real-time 360° horizontal field of view with its rotating head design. The rotation rate is user-selectable from 5–20 Hz to enable the user to determine the density of data points generated by the LiDAR sensor. The HDL-64 device generates laser with 905 nm wavelength and 5 ns pulse, which captures a point cloud of up to 2,200,000 points/s with a range of up to 120 m and a typical accuracy of ±2 cm. The upper part of the device consists of the laser emitters (4 groups of 16 each), and the lower part of the device consists of laser receivers (2 groups of 32 each).

In practice, one major problem faced by LiDAR manufacturers as well as users is calibration [21]. The performance of LiDAR devices strongly depends on their calibration. With good calibration, precise 3D data from an environment can easily be processed to extract linear or planar features. On the contrary, the extraction of these features can be difficult, unreliable or impossible if the sensor is badly calibrated.

A multi-beam LiDAR system is modeled as a set of rays, i.e., straight lines. These rays define the position and orientation of laser beams in a sensor-fixed coordinate frame. The intrinsic calibration for such systems is the estimation of parameters that define the position and orientation of each of the laser beams. The principle underlying the calibration techniques is an optimization process performed to estimate the LiDAR calibration parameters so that the 3D data acquired by LiDAR matches the ground truth. The calibration process is an optimization process that involves many parameters, and it can be divided into the following steps.

- **Choice of parameterization:** At least five parameters are required to define one laser beam in a 3D coordinate frame, including two angles to define the direction of the associated line and three parameters to define the point origin of the beam. If a distance correction factor is required to correct the measurement made by laser beam, the number of calibration parameters goes to six or seven per laser beam.

- **Choice of objective function:** An objective/cost function C forms the basis of the optimization process and is used to quantitatively compare the acquired 3D point cloud data and the real environment. C should provide higher costs if there is more difference between the acquired 3D data and ground-truth environment, and lower costs as the match between acquired 3D data and real environment improves.

- **Data segmentation:** This step consists in extracting, from acquired data, the data that actually correspond to the calibration object for which the ground truth is known. The environment chosen for the calibration process should be designed and made to allow appropriate segmentation of data.

As the number of beams increases, so as the number of calibration parameters. Therefore, the calibration process is more difficult for devices with higher number of beams. This is one major reason why devices with higher number of beams are much more expensive compared to devices with lower number of beams.

2.2.2   HIGH-DEFINITION MAPS OVERVIEW

In this subsection, we explore the technical details behind the making of HD maps. But first, why HD maps are needed for autonomous driving? Reflect for a moment on driving a very familiar route—from your home to your office, for example. You already have a mental "map" of your commute before you begin driving, making it easier to focus on the truly safety-critical parts of the drive. For instance, you can anticipate unusual driver avoidance behavior where a large pothole has been for weeks, and know the speed limit despite signs being blocked by a large truck. Now compare this to driving a completely new route, when you have much more information to process because everything is unfamiliar. You can only react to what you see in the moment. The same principles apply to autonomous vehicles. HD maps make routes familiar to autonomous vehicles, thus making them safer. Next, why not use existing digital maps for autonomous driving? Existing digital maps are meant to be consumed by humans, they usually have low resolutions (meter level precision) and not updated frequently. On the contrary, in order for an autonomous vehicle to follow a road, it needs to know where the road is. To stay in a specific lane, it needs to know where the lane is. For an autonomous vehicle to stay in a lane, the localization requirements are in the order of decimeters. Therefore, using existing digital maps, it is very hard for autonomous vehicles to perform accurate real-time localization, especially when the environment does not match what is on the map.

There are three critical challenges of map making for autonomous vehicles: maps need to be very precise (centimeter level precision), and hence HD; they need to be fresh if they are to reflect changes on the roads (in practice the refresh rate is once/wk); and they need to work seamlessly with the rest of autonomous driving system with high performance. To achieve high precision, we can utilize LiDAR in combination with other sensors to capture map data. To achieve map freshness, we can crowd-source the map-making process (the DeepMap approach) as opposed to having survey fleets generate maps periodically (the Google and Baidu approach). To have the HD maps seamlessly work with the rest of the autonomous driving system, we can build a high-performance autonomous driving cloud infrastructure to provide real-time HD map updates to autonomous vehicles. In the rest of this subsection, we focus on how to build HD maps with centimeter-level precision.

The core idea of HD map making is to augment GNSS/INS navigation by learning a detailed map of the environment, and then to use a vehicle's LiDAR sensor to localize the vehicle relative to the HD map [22, 23, 24]. Thus, the key of HD map making is to fuse different sensors information (GNSS/INS/LiDAR) to minimize the errors in each grid cell of the HD map. In this process, GNSS/INS first generates rough location information for each scan and then LiDAR provides high precision for each 2D location in the environment. The key challenges include how to perform sensor fusion to derive high-precision local maps and how to generate global maps by stitching the local maps together.

First let us study the anatomy of HD maps. Similar to traditional maps, HD maps also have hierarchical data structures. The foundation or bottom layer is a high-precision 2D with resolution of 5 × 5 cm [23]. This foundation layer captures 2D overhead views of the road surface, taken in the infrared spectrum with LiDAR. Each grid cell in this foundation layer stores the LiDAR reflectivity information in each grid cell. Through the reflectivity information we can judge whether a grid cell is a clear road surface or it is occupied by obstacles. As we will discuss later, to localize against this HD map, in real-time autonomous vehicles compare its current LiDAR scans against the LiDAR reflectance information that is stored in the foundation layer grid cells.

Figure 2.6: HD map. Courtesy of DeepMap, used with permission.

More details about the foundation layer, which we can treat as an orthographic infrared photograph of the ground, as each 2D grid cell is assigned an x-y location in the environment with an infrared reflectivity value. To capture the raw LiDAR scan, one or more LiDAR sensors are mounted on a vehicle, pointing downward at the road surface. In addition to returning the range to a sampling of points on the ground, these lasers also return a measure of infrared reflectivity. By texturing this reflectivity data onto the 3D range data, the result is a dense infrared reflectivity image of the ground surface. To eliminate the effect of non-stationary objects in the map on subsequent localization, one standard approach is to fit a ground plane to each laser scan, and only retains measurements that coincide with this ground plane [23]. The ability to remove vertical objects is a key advantage of using LiDAR sensors over conventional cameras. As a result, only the flat ground is mapped, and other vehicles are automatically discarded from the data. Maps like these can be acquired even at night, as the LiDAR system does not rely on external light. This makes the mapping result much less dependent on ambient lighting than is the case for passive cameras.

Once we have captured LiDAR scans, we can treat each scan as a local map of the environment. However, to generate a large-scale map, we need a way to stitch all the local maps together into a global map through a process called map matching [24]. Map matching compares local

LiDAR scans and identify regions of overlap among these local scans, and then using the over-lapped regions as anchors to stitch the maps together. Formally, let us define two LiDAR scan sequences as two disjoint sequences of time indices, $a1, a2, \ldots$ and $b1, b2, \ldots$, such that the corre-sponding grid cells in the map show an overlap that exceeds a given threshold T. Once such a region is found, two separate maps can be generated, one using only data from $a1, a2, \ldots$, and the other only with data from $b1, b2, \ldots$. It then searches for the alignment that maximizes the measurement probability, assuming that both adhere to a single maximum likelihood infrared reflectivity map in the area of overlap. Specifically, a linear correlation field is computed between these maps, for different x-y offsets between these images. Note that since we have GNSS and INS data when capturing the LiDAR scans, each LiDAR scan has been post-processed with GNS and INS data such that each LiDAR scan is associated with an initial pose $< x, y, \theta >$, where x and y represent the exact location of the vehicle when the scan is taken, and θ represents the heading direction of the vehicle when the scan is taken. This pose information can be used to bound the errors of the map matching process when we compute correlation coefficients from elements whose infrared reflectivity value is known in these two maps. In cases where the alignment is unique, we find a single peak in this correlation field. The peak of this correlation field is then assumed to be the best estimate for the local alignment.

On top of the foundation layer, the HD map contains layers of semantic information. As shown in the Figure 2.7, the layer above the foundation layer contains the location and characteris-tic information of the road marking line, and the corresponding lane characteristics. As the vehicle's sensors may not be reliable under different conditions, such as bad weathers, obstructions, and in-terferences from other vehicles, the lane information feature in the HD map can help autonomous vehicles accurately and reliably identify road lanes, and in real-time identify whether the adjacent lanes are safe. On top of the road lane layers, HD maps also have layers to provide road signs and traffic signals, etc. This layer provides two functions: as an input to the perception stage to have the car getting ready to detect traffic signs and speed limits; or as an input to the planning stage, such that the vehicle can still travel safely using the traffic signs and speed limits contained in the HD map in case the vehicle fails to detect these signs and limits.

Next problem faced by HD maps is the storage space, as HD results in high storage and memory space. Maps of large environments at 5-cm resolution occupy a significant amount of memory. As proposed in [23], two methods can be utilized to reduce the size of the maps and to allow relevant data to fit into main memory. The first method filters out irrelevant information: when acquiring data in a moving vehicle, the rectangular area which circumscribes the resulting laser scans grows quadratically with travelled distance, despite that the data itself grows only linearly. In order to avoid a quadratic space requirement, we can break the rectangular area into a square grid, and only save squares for which there is data. With this approach, the grid images require approximately 10 MB per mile of road at 5-cm resolution. This would allow a 1 TB hard drive to hold 100,000 miles of data. While the first method optimizes storage usage, the second method targets memory usage. At any moment as the vehicle travel, we only need a local HD map. Also, as we have GNSS/INS information to help us roughly locate the vehicle in real time, we can use this information to dynamically preload a small portion of the HD map into memory regardless of the size of the overall map.

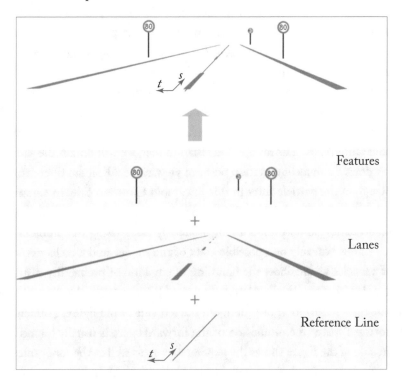

Figure 2.7: Semantic layers of HD maps.

2.2.3   LOCALIZATION WITH LIDAR AND HD MAP

Once the HD map has been built, we need to localize a vehicle in real-time against the HD map [25, 26, 27, 28]. The standard approach to achieve this is through particle filter, which analyzes range data in order to extract the ground plane underneath the vehicle. It then correlates via the Pearson product-moment correlation the measured infrared reflectivity with the map. Particles are projected forward through time via the velocity outputs from a tightly coupled inertial navigation system, which relies on wheel odometry, an INS and a GNSS system for determining vehicle velocity.

Before we dive into the details of localization against HD maps, let us take a moment to understand the mathematical tools used in localization. Before particle filtering methods became popular, Kalman filtering was the standard method for solving state space models. A Kalman Filter can be applied to optimally solve a linear Gaussian state space model. When the linearity or Gaussian conditions do not hold, its variants, the Extended Kalman Filter (EKF) and the Unscented Kalman Filter, can be used. However, for highly nonlinear and non-Gaussian problems, they fail to provide a reasonable estimate. Particle filtering techniques offer an alternative method. They work online to approximate the marginal distribution of the latent process as observations become available. Importance sampling is used at each time point to approximate the distribution with a set of discrete values, known as particles, each with a corresponding weight [29]. A Kalman filter relies on the linearity and normality assumptions. Sequential Monte Carlo methods, in particular particle filter, reproduce the work of the Kalman filter in those nonlinear and/or non-Gaussian environments. The key difference is that, instead of deriving analytic equations as a Kalman filter does, a particle filter uses simulation methods to generate estimates of the state and the innovations. If we apply particle filtering to a linear and Gaussian model, we will obtain the same likelihood as the Kalman filter does. From a computation point of view, since a Kalman filter avoids simulations, it is less expensive than the particle filter in this linear and Gaussian case. In summary, if a system does not fit nicely into a linear model, or the sensor's uncertainty does not look "very" Gaussian, a particle filter would handle almost any kind of model, by discretizing the problem into individual "particles"—each one is basically one possible state of the model, and a collection of a sufficiently large number of particles would allow the handling of any kind of probability distribution.

Applying a Particle filter, localization against the HD map takes place in real time. Figure 2.8 shows the particle filter localization in action. On the left side of the figure, a bunch of particles are thrown in the forward space (the dimension of the forward space is usually bound by GNSS/INS errors). The right side of the figure shows the generated map as well as the robot trace. Each particle in the forward space is associated with a weight, the higher the weight, the more likely it represents the vehicle's current location. In this particular example, the red particle represents higher weight whereas the black particle represents lower weight. The Particle filter algorithm is recursive in nature and operates in two phases: *prediction and update*. After each action, each particle is modified according to the existing model, including the addition of random noise in order to simulate the effect of noise on the variable of interest, this is the *prediction* stage. Then each particle's weight is re-evaluated based on the latest LiDAR scan, and this is the *update* stage. Specifically, if a LiDAR scan matches the environment around a particle (which is a point in the HD map), then it is likely that the particle is very close to the exact location of the vehicle, and thus it is assigned a very high weight. After the update stage, if a significant number of particles with high weight concentrate on a small area (a scenario called convergence), then we re-perform the whole process within a smaller region (a process called resampling) to further refine the localization accuracy.

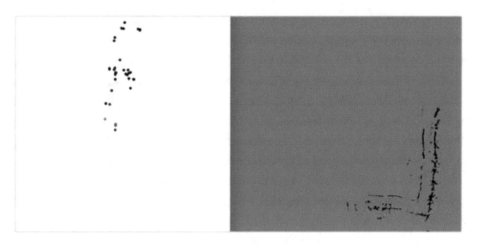

Figure 2.8: Particle filter in action.

To achieve real-time performance, the Particle filter maintains a three-dimensional pose vector (*x*, *y*, and *yaw*), whereas *roll* and *pitch* are assumed to be sufficiently accurate as is. The motion prediction in the particle filter is based on inertial velocity measurements. As in the mapping step, a local ground plane analysis removes LiDAR measurements that correspond to non-ground objects. Further, measurements are only incorporated for which the corresponding map is defined, that is, for which a prior reflectivity value is available in the map. To bound localization errors, a small number of particles are continuously drawn from current GNSS/INS pose estimate. GNSS, when available, is also used in the calculation of the measurement likelihood to reduce the danger of particles moving too far away from the GNSS location. One complicating factor in vehicle localization is weather. The appearance of the road surface is affected by rain, in that wet surfaces tend to reflect less infrared laser light than do dry ones. To adjust for this effect, the particle filter normalizes the brightness and standard deviation for each individual range scan, and also for the corresponding local map stripes. This normalization transforms the least squares difference method into the computation of the Pearson product-moment correlation with missing variables.

The previous paragraphs explain the basic approach of LiDAR-based localization against HD maps. Now we examine the problems in the basic approach. The first problem is localization error: the basic approach discussed in the previous paragraph uses a binary classification for deciding whether or not a LiDAR scan matches part of the HD map. If a scan is determined not a match, the LiDAR scan is discarded. While this approach is simple and clean, in practice it often leads to false positives or false negatives, resulting in high error rates. To improve upon this binary classification approach, in [25], the authors propose to extend the HD map to encapsulate the probabilistic nature of the environment, so as to represent the world more accurately and localize with fewer errors. In their approach, instead of having to explicitly decide whether each LiDAR scan either is or is not part of the static environment, the authors take into account the sum of all observed data and model the variances observed in each part of the map. This new approach has several advantages compared to the original non-probabilistic approach. First, although retro-reflective surfaces have the fortuitous property that remissions are relatively invariant to angle of incidence, angular-reflective surfaces such as shiny objects yield vastly different returns from different positions. Instead of ignoring these differences, which can lead to localization errors, this new approach implicitly accounts for them. Second, this approach provides an increased robustness to dynamic obstacles: by modeling distributions of reflectivity observations in the HD map, dynamic obstacles are automatically discounted in localization via their trails in the map. Third, this approach enables a straightforward probabilistic interpretation of the measurement model used in localization.

The second problem is the high cost of LiDAR sensors: 3D LiDAR devices are very expensive, costing around $100,000 per unit. This high cost can become a major blocker for the commercialization of autonomous vehicles. To address this problem, many different approaches of utilizing more cost-effective sensors have been proposed. Particularly, in [26], the authors demonstrate a method for precisely localizing a road vehicle using a single push-broom 2D laser scanner while leveraging a prior HD map (generated using high-end LiDAR device). In their setup, the 2D laser is oriented downward, thus causing continual ground strike such that they can produce a small 3D swathe of LiDAR data, which can be matched statistically against the HD map. For this technique to work, we need to provide accurate vehicle velocity information to the localization module at real time, and this velocity information can be obtained from vehicle speedometers. Using this approach, the authors managed to outperform an accurate GNSS/INS localization system. Similar approach has been proposed in [27] as well, where the authors present a localization algorithm for vehicles in 3D urban environment with only one 2D LiDAR and odometry information.

The third problem is weather conditions (e.g., rain, snow, mud): as LiDAR performance can be severely affected by weather conditions. For instance, when faced with adverse weather conditions that obscure the view of the road paint or poor road surface texture, LiDAR-based localization solution often fails. One interesting observation is that adverse weather conditions usually affect the reflectivities of the LiDAR scans, thus if we could use some other information instead of the reflectivities for localization, we could potentially solve this problem. Hence, in [28], the authors propose to use the 3D structure of the scene (z-height) instead of reflectivities for scan matching. To achieve this, the authors proposed to leverage Gaussian mixture maps to exploit the structure in the environment. These maps are a collection of Gaussian mixtures over the z-*height* distribution. To achieve real-time performance, the authors also develop a novel branch-and-bound, multi-resolution approach that makes use of rasterized lookup tables of these Gaussian mixtures.

2.3   VISUAL ODOMETRY

Visual odometry (VO) is the process of estimating the egomotion of a vehicle using only the input of one or more cameras [30, 31, 32]. The main task in VO is to compute the relative transformations T_x from the images I_x and I_{x-1} and then to utilize the transformations to recover the full trajectory $V_{0:n}$ of the vehicle. This means that VO recovers the path incrementally, pose after pose. An iterative refinement over the last x poses can be performed after this step to obtain a more accurate estimate of the local trajectory. This iterative refinement works by minimizing the sum of the squared re-projection errors of the reconstructed 3D points over the last m images, this process is commonly called *Bundle Adjustment*.

A typical VO pipeline is shown in Figure 2.9. For every new image I_x, the first step is to extract the feature points from the image; then, the second step is to match the extracted feature points with those from the previous frames. Note that 2D features that are the reprojection of the same 3D feature across different frames are called image correspondences. The third step is motion estimation, which consists of computing the relative motion T_x between the time instants $x-1$ and x. The vehicle pose V_x is then computed by concatenation of T_x with the previous pose. Finally, *Bundle Adjustment* can be done over the last x frames to obtain a more accurate estimate of the local trajectory.

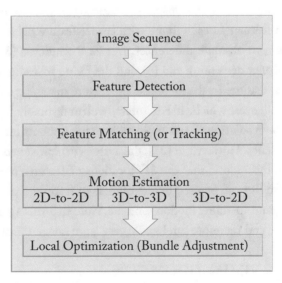

Figure 2.9: A typical VO pipeline (based on [30]).

Motion estimation is the core computation step performed for every image in a VO system. More precisely, in the motion estimation step, the vehicle motion between the current image and the previous image is computed. By combining all these motions, the full trajectory of the vehicle can be recovered. Depending on whether the feature correspondences are specified in two or three dimensions, there are three different methods. 2D-to-2D: In this case, both f_{x-1} and fx are specified in 2D image coordinates. 3D-to-3D: In this case, both f_{x-1} and f_x are specified in 3D. To achieve this, we first triangulate 3D points for each frame. 3D-to-2D: In this case, f_{k-1} are specified in 3D and f_k are their corresponding 2D reprojections on the image I_x. In the monocular case, the 3D structure needs to be triangulated from two adjacent images I_{x-2} and I_{x-1} and then matched to 2D image features in a third image I_x.

2.3.1   STEREO VISUAL ODOMETRY

Based on the observation that inertial sensors are prone to drift and wheel odometry is unreliable in bumpy "off-road" terrain, we can utilize visual odometry for real-time localization. Stereo visual odometry works by estimating frame-to-frame camera motion from successive stereo image pairs and it has been successfully applied to autonomous driving. For instance, the algorithm presented in [35] differs from most visual odometry algorithms in two key respects: (1) it makes no prior assumptions about camera motion, and (2) it operates on dense disparity images computed by a separate stereo algorithm. The authors have demonstrated that after 4,000 frames and 400 m of travel, position errors are typically less than 1 m (0.25% of distance traveled); and the processing time is only about 20 ms on a 512 × 384 image. The basic algorithm works as follows: for a given pair of frames, (1) detect features in each frame with corner feature detection; (2) match features between frames using sum-of-absolute differences over local windows; (3) find the largest set of self-consistent matches (inlier detection); and (4) find the frame-to-frame motion that minimizes the re-projection error for features in the inlier set. The feature matching stage inevitably produces some incorrect correspondences, which, if left intact, will unfavorably bias the frame-to-frame motion estimate. A common solution to this problem is to use a robust estimator that can tolerate some number of false matches. However, in [35], the authors exploit stereo range data at the inlier detection stage. The core intuition is that the 3D locations of features must obey a rigidity constraint, and that this constraint can be used to identify sets of features that are mutually consistent prior to computing the frame-to-frame motion estimate. Note that this algorithm does not require an initial motion estimate, and therefore can handle very large image translations.

2.3.2   MONOCULAR VISUAL ODOMETRY

The difference from the stereo scheme is that in the monocular VO, both the relative motion and 3D structure must be computed from 2D data since the absolute scale is unknown. The distance between the first two camera poses is usually set to fixed distance in order to establish initial 3D structure. As a new image arrives, the relative scale and camera pose with respect to the first two frames are determined using the knowledge of 3D structure. Monocular VO can be divided into three categories: feature-based methods, appearance-based methods, and hybrid methods. Feature-based methods are based on salient and repeatable features that are tracked over the frames; appearance-based methods use the intensity information of all the pixels in the image or sub-regions of it; and hybrid methods use a combination of the previous two.

In [36], the authors proposed an autonomous driving localization system based on mono omnidirectional camera. They indicated the necessity to use omnidirectional cameras for the following reasons: (1) many outliers, such as moving vehicles, oscillating trees, and pedestrians, are present; (2) at times very few visual features are available; (3) occlusions, especially from the trees, make it almost impossible to track landmarks for a long period of time. The proposed system is a fully incremental localization system that precisely estimates the camera trajectory without relying on any motion model. Using this algorithm, at a given time frame, only the current location is estimated while the previous camera positions are never modified. The key of the system is a fast and simple pose estimation algorithm that uses information not only from the estimated 3D map, but also from the epipolar constraint. The authors demonstrated that using epipolar constraints leads to a much more stable estimation of the camera trajectory than the conventional approach.

To verify the effectiveness of the proposed algorithm, the authors conducted an experiment to have the vehicle travel 2.5 km, during which the frame rate was set to 10 images per second, and the results show that with this system, the localization error could be controlled to around 2%.

2.3.3   VISUAL INERTIAL ODOMETRY

Inertial sensors provide very frequent updates (1 KHz) although they are subject to drift problems. On the other hand, although visual odometry provides accurate position updates, when the vehicle makes sharp turns VO often loses track of its position due to the lack of matched feature points caused by infrequent image updates. Since visual and inertial measurements offer complementary properties, they are particularly suitable for fusion to provide robust and accurate localization and mapping, a primary need for any autonomous vehicle system. The technique of fusing visual and inertial sensors for real-time localization is called Visual Inertial Odometry (VIO). There are two main concepts toward approaching the visual-inertial estimation problem: batch nonlinear optimization methods [37] and recursive filtering methods [38]. The former jointly minimizes the error originating from integrated inertial measurements and the visual reprojection errors from visual, whereas the latter utilizes the inertial measurements for state propagation while updates originate from the visual observations.

In [37] the authors incorporate inertial measurements into batch visual SLAM. A nonlinear optimization is formulated to find the camera poses and landmark positions by minimizing the reprojection error of landmarks observed in camera frames. As soon as inertial measurements are introduced, they not only create temporal constraints between successive poses, but also between successive speed and inertial sensor bias estimates of both accelerometers and gyroscopes by which the robot state vector is augmented. The authors formulate the visual-inertial localization and mapping problem as one joint optimization of a cost function $J(x)$ containing both the visual reprojection errors and the temporal error term from the inertial sensor, where x represents the current state of the vehicle.

The first step is propagation. At the beginning, inertial measurements are used to propagate the vehicle pose in order to obtain a preliminary uncertain estimate of the states. Assume a set of past frames as well as a local map consisting of landmarks with sufficiently well-known 3D position is available at this point. As a first stage of establishing correspondences, a 3D-2D matching step is performed. Then, 2D-2D matching is performed in order to associate key points without 3D landmark correspondences. Next, triangulation is performed to initialize new 3D landmark positions. Both stereo-triangulation across stereo image pairs as well as between the current frame and any previous frame available is performed.

The second step is optimization, when a new image frame comes in, features are extracted from the image to extend new 3D points, which can be used to extend the map as discussed above. Once in a while, a frame is selected as a key frame, which triggers optimization. One simple heuristic to select key frame is that if the ratio of matched vs. newly detected feature points is small, the frame is inserted as keyframe. Then the new key frame, along with all previous key frames, are used in a global optimization to minimize both the visual reprojection errors and the inertial sensor temporal error. The experiment results show that, with this optimization method, with a 500 m travel, the translation error is less than 0.3%.

Although the optimization methods usually provide higher localization accuracy, the need for multiple iterations also incurs a higher computational cost. A lightweight approach, based on the Extended Kalman Filter (EKF) has been proposed for vehicle localization [38]. In this approach, the authors use EKF algorithms to maintain a sliding window of camera poses in the state vector, and use the feature observations to apply probabilistic constraints between these poses. The algorithm can be divided into the following steps: (a) *Propagation:* for each inertial measurement, propagate the filter state and covariance; (b) *Image registration*: every time a new image is recorded, augment the state and covariance matrix with a copy of the current camera pose estimate; and (c) *Update:* when the feature measurements of a given image become available, perform an EKF update. Using this approach, the authors performed an experiment to have a vehicle travel 3.2 km, and the final position error was only 10 m, an error of only 0.31% of the traveled distance.

2.4   DEAD RECKONING AND WHEEL ODOMETRY

Dead reckoning (derived from "deduced reckoning" of sailing days) is a simple mathematical procedure for determining the present location of a vehicle by advancing some previous position through known course and velocity information over a given length of time [39]. The vast majority of autonomous vehicle systems in use today rely on dead reckoning to form the very backbone of their navigation strategy. The most simplistic implementation of dead reckoning is wheel odometry, or deriving the vehicle displacement along the path of travel from wheel encoders. In this section, we study wheel encoders, sources of wheel odometry errors, as well as methods to reduce wheel odometry errors.

2.4.1   WHEEL ENCODERS

A common means of odometry instrumentation involves optical encoders directly coupled to the motor armatures or wheel axles. Since most mobile robots rely on some variation of wheeled locomotion, a basic understanding of sensors that accurately quantify angular position and velocity is an important prerequisite to further discussions of odometry. There are many different types of wheel encoders, including *optical encoders, Doppler encoders, differential drive, tricycle drive, Ackerman Steering, Synchro drive, Omnidirectional drive, racked vehicle*, etc.

Since Ackerman Steering (AS) is used almost exclusively in the automotive industry, we focus on AS in this sub-section. AS is designed to ensure that the inside front wheel is rotated to a slightly sharper angle than the outside wheel when turning, thereby eliminating geometrically induced tire slippage. As shown in Figure 2.10, the extended axes for the two front wheels intersect in a common point that lies on the extended axis of the rear axle. The locus of points traced along the ground by the center of each tire is thus a set of concentric arcs about this center-point of rotation P_1, and all instantaneous velocity vectors will subsequently be tangential to these arcs. Such a steering geometry is said to satisfy the Ackerman equation:

$$\cot \theta_i = \cot \theta_o = \frac{d}{l},$$

where θ_i represents the relative steering angle of the inner wheel, and θ_o represents relative steering angle of the outer wheel, l represents longitudinal wheel separation, and d represents lateral wheel separation.

AS provides a fairly accurate odometry solution while supporting the traction and ground clearance needs of all-terrain operation. AS is thus the method of choice for outdoor autonomous vehicles. AS implementations typically employ a gasoline or diesel engine coupled to a manual or automatic transmission, with power applied to four wheels through a transfer case, a differential, and a series of universal joints.

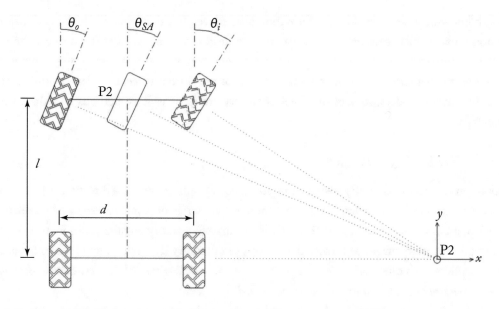

Figure 2.10: An Ackerman-steered vehicle (based on [39]).

2.4.2   WHEEL ODOMETRY ERRORS

As indicated in [39], wheel odometry uses encoders to measure wheel rotation and/or steering orientation. It is well known that odometry provides good short-term accuracy, is inexpensive, and allows very high sampling rates. However, the fundamental idea of odometry is the integration of incremental motion information over time, which leads inevitably to the accumulation of errors. Particularly, the accumulation of orientation errors will cause large position errors that increase proportionally with the distance traveled by the robot.

Odometry data can be fused with absolute position measurements to provide better and more reliable position estimation. Odometry can be used in between absolute position updates with landmarks (such as visual landmarks/features used in visual odometry). Given a required positioning accuracy, increased accuracy in odometry allows for less frequent absolute position updates. As a result, fewer landmarks are needed for a given travel distance. Many mapping and landmark matching algorithms assume that the vehicle can maintain its position well enough to allow the vehicle to look for landmarks in a limited area and to match features in that limited area to achieve short processing time and to improve matching correctness. In some cases, odometry is the only navigation information available; for example: when no external reference is available, when circumstances preclude the placing or selection of landmarks in the environment, or when another sensor subsystem fails to provide usable data.

Odometry is based on the assumption that wheel revolutions can be translated into linear displacement relative to the floor. This assumption is only of limited validity. One extreme example is wheel slippage: if one wheel was to slip on, say, an oil spill, then the associated encoder would register wheel revolutions even though these revolutions would not correspond to a linear displacement of the wheel. The wheel odometry error sources fit into one of two categories: systematic errors and non-systematic errors. Systematic errors include *unequal wheel diameters, average of actual wheel diameters differs from nominal wheel diameter, actual wheelbase differs from nominal wheelbase, misalignment of wheels, finite encoder resolution,* as well as *finite encoder sampling rate.* Non-systematic errors include *travel over uneven floors, travel over unexpected objects on the floor, wheel-slippage due to slippery floors, wheel-slippage due to over-acceleration, fast turning* (skidding), *external forces* (interaction with external bodies), *internal forces* (castor wheels), as well as *non-point wheel contact with the floor.*

The clear distinction between systematic and non-systematic errors is of great importance for the effective reduction of odometry errors. For example, systematic errors are particularly grave because they accumulate constantly. On most smooth indoor surfaces systematic errors contribute much more to odometry errors than non-systematic errors. However, on rough surfaces with significant irregularities, non-systematic errors are dominant. The problem with non-systematic errors is that they may appear unexpectedly and they can cause large position errors. Typically, when an autonomous vehicle system is installed with a hybrid wheel odometry/vision navigation system, the frequency of the images is based on the worst-case systematic errors. Such systems are likely to fail when one or more large non-systematic errors occur.

2.4.3   REDUCTION OF WHEEL ODOMETRY ERRORS

The accuracy of wheel odometry depends to some degree on their kinematic design and on certain critical dimensions. Here we first summarize some of the design-specific considerations that affect dead-reckoning accuracy [39]: Vehicles with a small wheelbase are more prone to orientation errors than vehicles with a larger wheelbase. Vehicles with castor wheels that bear a significant portion of the overall weight are likely to induce slippage when reversing direction (the "shopping cart effect"). Conversely, if the castor wheels bear only a small portion of the overall weight, then slippage will not occur when reversing direction. In addition, it is widely known that, ideally, wheels used for odometry should be "knife-edge" thin and not compressible. The ideal wheel would be made of aluminum with a thin layer of rubber for better traction. In practice, however, this design is not feasible because the odometry wheels are usually also load-bearing drive wheels, which require a somewhat larger ground contact surface. In the rest of this sub-section we summarize methods for reducing systematic and non-systematic wheel odometry errors.

Methods for reducing systematic odometry errors include the following.

- **Auxiliary Wheels and Basic Encoder Trailer:** It is generally possible to improve odometric accuracy by adding a pair of "knife-edge," non-load-bearing encoder wheels. Since these wheels are not used for transmitting power, they can be made to be very thin and with only a thin layer of rubber as a tire. Such a design is feasible for differential-drive, tricycle-drive, and Ackerman vehicles.

- **The Basic Encoder Trailer:** An alternative approach is the use of a trailer with two encoder wheels. It is virtually impossible to use odometry with tracked vehicles, because of the large amount of slippage between the tracks and the floor during turning. The idea of the encoder trailer is to perform odometry whenever the ground characteristics allow one to do so. Then, when the vehicle has to move over small obstacles, stairs, or otherwise uneven ground, the encoder trailer would be raised. The argument for this part-time deployment of the encoder trailer is that in many applications the vehicle may travel mostly on reasonably smooth concrete floors and that it would thus benefit most of the time from the encoder trailer's odometry.

- **Systematic Calibration:** Another approach to improving odometric accuracy without any additional devices or sensors is based on the careful calibration of a vehicle. As systematic errors are inherent properties of each individual vehicle. They change very slowly as the result of wear or of different load distributions. Thus, these errors remain almost constant over extended periods of time. One way to reduce such errors is vehicle- specific calibration. Nonetheless, calibration is difficult because even minute deviations in the geometry of the vehicle or its parts may cause substantial odometry errors.

Methods for reducing non-systematic odometry errors include the following.

- **Mutual Referencing:** We can use two robots that could measure their positions mutually. When one of the robots moves to another place, the other remains still, observes the motion, and determines the first robot's new position. In other words, at any time one robot localizes itself with reference to a fixed object: the standing robot. However, this stop and go approach limits the efficiency of the robots.

- **Internal Position Error Correction (IPEC):** With this approach two mobile robots mutually correct their odometry errors. However, unlike the *mutual referencing* approach, the IPEC method works while both robots are in continuous, fast motion. To implement this method, it is required that both robots can measure their relative distance and bearing continuously and accurately. The principle of operation is based on the concept of error growth rate, which differentiates "fast-growing" and "slow-growing" odometry errors. For example, when a differentially steered robot traverses a floor irregularity it will immediately experience an appreciable orientation error (i.e., a fast-growing error). The associated lateral displacement error, however, is initially very small (i.e., a slow-growing error), but grows in an unbounded fashion as a consequence of the orientation error. The internal error correction algorithm performs relative position measurements with a sufficiently fast update rate (20 ms) to allow each truck to detect fast-growing errors in orientation, while relying on the fact that the lateral position errors accrued by both platforms during the sampling interval were small.

2.5   SENSOR FUSION

In the previous sections we have introduced different localization techniques. In practice, in order to achieve robustness and reliability, we often utilize sensor fusion strategy to combine multiple sensors together for localization [40, 41, 42]. In this section, we introduce three real-world examples of autonomous vehicles and study their localization approaches.

2.5.1   CMU BOSS FOR URBAN CHALLENGE

Figure 2.11: CMU autonomous vehicle [47].

As shown in Figure 2.11, Boss is an autonomous vehicle that uses multiple on-board sensors (GPS, LiDARs, radars, and cameras) to track other vehicles, detect static obstacles, and localize itself relative to a road model [40], and it is capable of driving safely in traffic at speeds up to 48 km/h. The system was developed from the ground up to address the requirements of the DARPA Urban Challenge. In this subsection we study Boss's localization system.

Boss's localization process starts with a differential GPS-based pose estimation. To do this it combines data from a commercially available position estimation system and measurements of road lane markers with an annotated road map. The initial global position estimate is produced by a sensor fusion system that combines differential GPS, IMU, and wheel encoder data to provide a 100-Hz position estimate, which is robust to GPS dropout. With a stable GPS signal, this system bounds the localization error to within 0.3 m; and thanks to sensor fusion, after 1 km of travel without GPS signal, it can still bound the localization error to within 1 m.

Although this system provides very high accuracy, it does not provide lane information. To detect lane boundaries, down-looking SICK LMS LiDARs are used to detect the painted lane markers on roads. Lane markers are generally brighter than the surrounding road material and therefore can be easily detected by convolving the intensities across a line scan with a slope function. Peaks and troughs in the response represent the edges of potential lane marker boundaries. To reduce false positives, only appropriately spaced pairs of peaks and troughs are considered to be lane markers. Candidate markers are then further filtered based on their brightness relative to their support region. The result is a set of potential lane marker positions.

To further improve localization accuracy, a road map is constructed to record static obstacle and lane information. Note that this road map is a precursor of the HD map. It basically extends a digital map with sub-meter accuracy with geometric features detected by LiDARs. These geometric features include obstacles and lane markers. The mapping system combines data from the numerous scanning lasers on the vehicle to generate both instantaneous and temporally filtered obstacle maps. The instantaneous obstacle map is used in the validation of moving obstacle hypotheses. The temporally filtered maps are processed to remove moving obstacles and are filtered to reduce the number of spurious obstacles appearing in the maps. Geometric features (curbs, berms, and bushes) provide one source of information for determining road shape in urban and off-road environments. Dense LiDAR data provide sufficient information to generate accurate, long-range detection of these relevant geometric features. Algorithms to detect these features must be robust to the variation in features found across the many variants of curbs, berms, ditches, embankments, etc. For instance, to detect curbs, the Boss team exploits two principle insights into the LiDAR data to simplify detection. First, the road surface is assumed to be relatively flat and slow changing, with road edges defined by observable changes in geometry, specifically in height. This simplification means that the primary feature of a road edge reduces to changes in the height of the ground surface. Second, each LiDAR scan is processed independently, as opposed to building a 3D point cloud. This simplifies the algorithm to consider input data along a single dimension. Then to localize the vehicle against the road map, a particle filter approach (as introduced in Section 2.3) can be utilized.

2.5.2   STANFORD JUNIOR FOR URBAN CHALLENGE

As shown in Figure 2.12, Junior is Stanford's entry in the Urban Challenge. Junior is a modified 2006 Volkswagen Passat Wagon, equipped with five LiDARs, a differential GPS-aided inertial navigation system, five radars, and two Intel quad core computer systems [41]. The vehicle has an obstacle detection range of up to 120 m, and reaches a maximum velocity of 48 km/h, which is the maximum speed limit according to the Urban Challenge rules.

Figure 2.12: Stanford autonomous vehicle [48].

Like the CMU Boss system, localization in Junior starts with a differential GPS-aided inertial navigation system, which provides real-time integration of GPS coordinates, inertial measurements, and wheel odometry readings. The real-time position and orientation errors of this system were typically below 100 cm and 0.1°, respectively. On top of this system, there are multiple LiDAR sensors providing real-time measurements of the adjacent 3D road structure as well as infrared reflectivity measurements of the road surface for lane marking detection and precision vehicle localization.

The vehicle is given a digital map of the road network. With the provided digital map and only the GPS-based inertial positioning system, Junior is not able to recover the coordinates of the vehicle with sufficient accuracy to perform reliable lane keeping without sensor feedback. Further, the digital map is itself inaccurate, adding further errors if the vehicle were to blindly follow the road. As a result, on top of the digital map, Junior performs fine-grained localization using local LiDAR sensor measurements. This fine-grained localization uses two types of information: road reflectivity and curb-like obstacles. The reflectivity is sensed using the RIEGL LMS-Q120 and the SICK LMS sensors, both of which are pointed toward the ground. To perform fine-grained localization, a 1D histogram filter is utilized to estimate the vehicle's lateral offset relative to the digital map. This filter estimates the posterior distribution of any lateral offset based on the reflectivity and the sighted curbs along the road. It rewards, in a probabilistic fashion, offsets for which lane-marker-like reflectivity patterns align with the lane markers or the road side in the digital map. The filter penalizes offsets for an observed curb would reach into the driving corridor of the digital map. As a result, at any point in time the vehicle estimates a fine-grained offset to the measured location by the GPS-based INS system.

2.5.3   BERTHA FROM MERCEDES BENZ

As shown in the previous sub-sections, both CMU Boss and Stanford Junior rely on GNSS/INS systems for coarse-grained localization, and then utilize local LiDAR scans and extended digital maps for fine-grained localization. However, LiDAR-based approach suffers from two major drawbacks: first, LiDAR devices are very expensive, costing over $80,000 USD per unit; second, it is even more expensive to build the HD map (millions of USD to build and maintain the HD map for each city). An alternative approach is to use computer vision and digital maps. One great example of this approach is the Bertha autonomous vehicle from Mercedes Benz, which relies solely on vision and radar sensors in combination with accurate digital maps to obtain a comprehensive understanding of complex traffic situations [42]. In this subsection, we study the localization details of the Bertha autonomous vehicle.

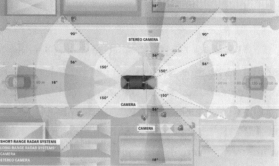

Figure 2.13: Mercedes Benz Bertha autonomous vehicle [42], used with permission.

As shown in Figure 2.13, the sensors used in this system are as follows: A GPS module is used for basic localization. Four 120° short-range radars are used for intersection monitoring. Two long-range radar mounted to the sides of the vehicle are used to monitor fast traffic at intersections on rural roads. A stereo camera system with 35 cm baseline for depth information at a range of 60 m. A wide-angle monocular color camera was mounted on the dash-board for traffic light recognition and pedestrian recognition in turning maneuvers. A wide-angle camera looking backwards is used for self-localization.

Another important source of information is a detailed digital map, which contains the position of lanes, the topology between them as well as attributes and relations defining traffic regulations, including but not limited to right-of-way, relevant traffic lights, and speed limits. The map used in the Bertha autonomous vehicle is created based on imagery from a stereo camera: For each stereo image pair, a dense disparity image and a 3D reconstruction of the vehicle's close environment are computed. These 3D points are projected onto the world plane and accumulated based on a reference trajectory. To ensure congruency, the same stereo images are also used for extracting the point feature map and the map containing visible lane markings. The reference trajectory is recorded by a RTK GNSS/INS navigation system whereas online localization during autonomous driving does not require such a costly system. For mapping and map maintenance, the Bertha team employed tools from the OpenStreetMap project [43].

To localize the Bertha autonomous vehicle, the GNSS/INS system typically bounds the localization error within a 1 m radius. On top of that, two complementary map relative localization algorithms were developed to further improve the localization accuracy to decimeter range. The first system detects point-shaped landmarks in the immediate vicinity of the vehicle and is specifically effective in urban areas with large man-made structures; this is called the feature-based localization. The other system exploits lane markings and curbstones as these are reliably detectable in rural areas and translates observations of these objects into a map-relative localization estimate, this is called the lane-marking-based localization. Note that both of these approaches utilize Visual Odometry techniques.

In the landmark-based approach, first a stereo image has been captured during the mapping stage, then during an autonomous test drive another image is captured from a rear facing monocular camera. The two images are registered spatially by means of a descriptor-based point feature association: salient features of the map sequence are associated with detected features in the current image of the vehicle's rear facing camera. Given the 3D positions of these landmarks have been reconstructed from the stereo image map sequence, it is possible to compute a 6D rigid-body transformation between both camera poses that would bring associated features in agreement. Fusing this transformation with the global reference pose of the map image and the motion information from wheel encoders and yaw rate sensors available in the vehicle, an accurate global position estimate can be recovered.

In rural areas often the only static features along the road are the markings on the road itself. Thus, a lane-marking-based localization system was developed. In a first step, a precise map containing all visible markings is built. In addition to the road markings and stop lines, also curbs and tram rails are annotated in the map. For the online localization step, a local section of this map is projected into the current image where the matching is done with a nearest neighbor search on the sampled map and the resulting residuals are minimized iteratively using a Kalman filter.

2.6   REFERENCES

[1]   Misra, P. and Enge, P., 2006. *Global Positioning System: Signals, Measurements and Performance,* 2nd ed. MA:Ganga-Jamuna Press. 15

[2]   Leick, A., Rapoport, L., and Tatarnikov, D. 2015. *GPS Satellite Surveying.* John Wiley & Sons. DOI: 10.1002/9781119018612. 16

[3]   Jeffrey, C. 2010. *An Introduction to GNSS.* Calgary, AB: NovAtel Inc. 17

[4]   Groves, P.D. 2013. *Principles of GNSS, Inertial, and Multisensor Integrated Navigation Systems.* Artech House. 17

[5]   Irsigler, M., Avila-Rodriguez, J.A., and Hein, G.W. 2005, September. Criteria for GNSS multipath performance assessment. In *Proceedings of the ION GNSS.* 17

[6]   Rieder, M.J. and Kirchengast, G. 2001. Error analysis and characterization of atmospheric profiles retrieved from GNSS occultation data. *Journal of Geophysical Research: Atmospheres*, 106(D23), pp. 31755–31770. DOI: 10.1029/2000JD000052. 17

[7]   Zumberge, J.F., Heflin, M.B., Jefferson, D.C., Watkins, M.M., and Webb, F.H. 1997. Precise point positioning for the efficient and robust analysis of GPS data from large networks. *Journal of Geophysical Research: Solid Earth*, 102(B3), pp. 5005–5017. DOI: 10.1029/96JB03860. 20

[8]   Gao, Y. and Chen, K. 2004. Performance analysis of precise point positioning using real-time orbit and clock products. *Journal of Global Positioning Systems*, 3(1-2), pp. 95–100. DOI: 10.5081/jgps.3.1.95. 20

[9]   Collins, J.P. and Langley, R.B. 1997. *A Tropospheric Delay Model for the User of the Wide Area Augmentation System.* Department of Geodesy and Geomatics Engineering, University of New Brunswick. 20

[10]   Wübbena, G., Schmitz, M., and Bagge, A. 2005, September. PPP-RTK: precise point positioning using state-space representation in RTK networks. In *Proceedings of ION GNSS* (Vol. 5, pp. 13–16). 20

[11]   Geng, J., Teferle, F.N., Meng, X., and Dodson, A.H. 2011. Toward PPP-RTK: Ambiguity resolution in real-time precise point positioning. *Advances in Space Research*, 47(10), pp. 1664–1673. DOI: 10.1016/j.asr.2010.03.030. 21

[12] Laurichesse, D. 2011, September. The CNES Real-time PPP with undifferenced integer ambiguity resolution demonstrator. In *Proceedings of the ION GNSS* (pp. 654–662). 21

[13] Grinter, T. and Roberts, C. 2013. Real time precise point positioning: Are we there yet? *IGNSS Symposium*, Outrigger Gold Coast, 16–18 July 2013, Paper 8. 21

[14] Caron, F., Duflos, E., Pomorski, D., and Vanheeghe, P. 2006. GPS/IMU data fusion using multisensor Kalman filtering: introduction of contextual aspects. *Information Fusion*, 7(2), pp. 221–230. DOI: 10.1016/j.inffus.2004.07.002. 21

[15] Schwarz, B. 2010. LIDAR: Mapping the world in 3D. *Nature Photonics,* 4(7), p. 429. DOI: 10.1038/nphoton.2010.148. 23

[16] Skinner, W.R. and Hays, P.B. 1994. A comparative study of coherent & incoherent Doppler lidar techniques. Marshall Space Flight Center. 23

[17] Spinhirne, J.D. 1993. Micro pulse lidar. *IEEE Transactions on Geoscience and Remote Sensing*, 31(1), pp. 48–55. DOI: 10.1109/36.210443. 23

[18] Baltsavias, E.P. 1999. Airborne laser scanning: existing systems and firms and other resources. *ISPRS Journal of Photogrammetry and Remote Sensing*, 54(2), pp. 164–198. DOI: 10.1016/S0924-2716(99)00016-7. 24

[19] Wehr, A. and Lohr, U. 1999. Airborne laser scanning—an introduction and overview. *ISPRS Journal of Photogrammetry and Remote Sensing*, 54(2), pp. 68–82. DOI: 10.1016/S0924-2716(99)00011-8. 24

[20] Donati, S. 1999. *Photodetectors* (Vol. 1). Prentice Hall PTR. 24

[21] Muhammad, N. and Lacroix, S. 2010, October. Calibration of a rotating multi-beam lidar. In *2010 IEEE/RSJ International Conference on Intelligent Robots and Systems (IROS)*, (pp. 5648–5653). IEEE. DOI: 10.1109/IROS.2010.5651382. 25

[22] Montemerlo, M. and Thrun, S. 2006. Large-scale robotic 3-d mapping of urban structures. *Experimental Robotics IX*, pp. 141–150. DOI: 10.1007/11552246_14. 26

[23] Levinson, J., Montemerlo, M., and Thrun, S. 2007, June. Map-based precision vehicle localization in urban environments. In *Robotics: Science and Systems* (Vol. 4, p. 1). DOI: 10.15607/RSS.2007.III.016. 26, 27, 28

[24] Konolige, K. 2004, July. Large-scale map-making. In *AAAI* (pp. 457–463). 26, 27

[25] Levinson, J. and Thrun, S. 2010, May. Robust vehicle localization in urban environments using probabilistic maps. In *2010 IEEE/RSJ International Conference on Robotics and Automation (ICRA)*, (pp. 4372–4378). IEEE. DOI: 10.1109/ROBOT.2010.5509700. 29, 31

[26]   Baldwin, I. and Newman, P. 2012, May. Road vehicle localization with 2d push-broom lidar and 3d priors. In *2012 IEEE International Conference on Robotics and Automation (ICRA)*, (pp. 2611–2617). IEEE. DOI: 10.1109/ICRA.2012.6224996. 29, 32

[27]   Chong, Z.J., Qin, B., Bandyopadhyay, T., Ang, M.H., Frazzoli, E., and Rus, D. 2013, May. Synthetic 2d lidar for precise vehicle localization in 3d urban environment. In *2013 IEEE International Conference on Robotics and Automation* (ICRA), (pp. 1554–1559). IEEE. DOI: 10.1109/ICRA.2013.6630777. 29, 32

[28]   Wolcott, R.W. and Eustice, R.M. 2015, May. Fast LIDAR localization using multiresolution Gaussian mixture maps. In *2015 IEEE International Conference on Robotics and Automation (ICRA)*, (pp. 2814–2821). IEEE. DOI: 10.1109/ICRA.2015.7139582. 29

[29]   Chen, Z. 2003. Bayesian filtering: From Kalman filters to particle filters, and beyond. *Statistics*, 182(1), pp. 1–69. DOI: 10.1080/02331880309257. 30

[30]   Scaramuzza, D. and Friedrich F. 2011. Visual odometry [tutorial]. *Robotics & Automation Magazine, IEEE* 18.4, pp. 80–92. DOI: 10.1109/MRA.2011.943233. 33

[31]   Nistér, D., Naroditsky, O., and Bergen, J. 2006. Visual odometry for ground vehicle applications. *Journal of Field Robotics*, 23(1), pp.3–20. DOI: 10.1002/rob.20103. 33

[32]   Forster, C., Pizzoli, M., and Scaramuzza, D. 2014, May. SVO: Fast semi-direct monocular visual odometry. In *2014 IEEE International Conference on Robotics and Automation (ICRA)*, (pp. 15–22). IEEE. DOI: 10.1109/ICRA.2014.6906584. 33

[33]   Tardif, J.P., Pavlidis, Y., and Daniilidis, K. 2008. September. Monocular visual odometry in urban environments using an omnidirectional camera. In *IROS 2008. IEEE/RSJ International Conference on Intelligent Robots and Systems, 2008*. (pp. 2531–2538). IEEE.

[34]   Scaramuzza, D., Fraundorfer, F., and Siegwart, R. 2009. May. Real-time monocular visual odometry for on-road vehicles with 1-point ransac. In *ICRA'09. IEEE International Conference on Robotics and Automation, 2009*. (pp. 4293–4299). IEEE.

[35]   Howard, A. 2008. September. Real-time stereo visual odometry for autonomous ground vehicles. In *Intelligent Robots and Systems, 2008*. IROS 2008. *IEEE/RSJ International Conference* on (pp. 3946–3952). IEEE. 34

[36]   Tardif, J.P., Pavlidis, Y., and Daniilidis, K. 2008, September. Monocular visual odometry in urban environments using an omnidirectional camera. In *IROS 2008. IEEE/RSJ International Conference on Intelligent Robots and Systems, 2008*. (pp. 2531–2538). IEEE. DOI: 10.1109/IROS.2008.4651205. 35

[37] Leutenegger, S., Lynen, S., Bosse, M., Siegwart, R., and Furgale, P. 2015. Keyframe-based visual–inertial odometry using nonlinear optimization. *The International Journal of Robotics Research*, 34(3), pp. 314–334. DOI: 10.1177/0278364914554813. 35

[38] Mourikis, A. I. and Roumeliotis, S. I. 2007. A multi-state constraint Kalman filter for vision-aided inertial navigation. In *Proceedings of the IEEE International Conference on Robotics and Automation (ICRA)*. DOI: 10.1109/ROBOT.2007.364024. 35, 36

[39] Borenstein, J., Everett, H.R., and Feng, L. 1996. *Where am I? Sensors and Methods for Mobile Robot Positioning*. University of Michigan, 119(120). 37, 38, 39

[40] Urmson, C., Anhalt, J., Bagnell, D., Baker, C., Bittner, R., Clark, M.N., Dolan, J., Duggins, D., Galatali, T., Geyer, C., and Gittleman, M. 2008. Autonomous driving in urban environments: Boss and the urban challenge. *Journal of Field Robotics*, 25(8), pp. 425–466. DOI: 10.1002/rob.20255. 41

[41] Montemerlo, M., Becker, J., Bhat, S., Dahlkamp, H., Dolgov, D., Ettinger, S., Haehnel, D., Hilden, T., Hoffmann, G., Huhnke, B., and Johnston, D. 2008. Junior: The stanford entry in the urban challenge. *Journal of Field Robotics*, 25(9), pp. 569–597. DOI: 10.1002/rob.20258. 41, 43

[42] Ziegler, J., Bender, P., Schreiber, M., Lategahn, H., Strauss, T., Stiller, C., Dang, T., Franke, U., Appenrodt, N., Keller, C.G., and Kaus, E. 2014. Making Bertha drive—An autonomous journey on a historic route. *IEEE Intelligent Transportation Systems Magazine*, 6(2), pp. 8–20. DOI: 10.1109/MITS.2014.2306552. 41, 44

[43] Haklay, M. and Weber, P. 2008. Openstreetmap: User-generated street maps. *IEEE Pervasive Computing*, 7(4), pp. 12–18. DOI: 10.1109/MPRV.2008.80. 45

[44] Van Sickle, J. 2015. *GPS for Land Surveyors*, 4th ed. CRC Press.

[45] Jurvetson, S. 2009. Velodyne High-Def LIDAR. Copyright 2009. CC BY 2.0. https://www.flickr.com/photos/44124348109@N01/4042817787. 24

[46] Van Sickle, J. 2001. *GPS for Land Surveyors*, 2nd ed. CRC Press. 19

[47] Jurvetson, S. 2008. CMU carbot. Copyright 2008. CC BY 2.0. https://www.flickr.com/photos/jurvetson/2328347458. 41

[48] Jurvetson, S. 2009. Hands-free Driving. Copyright 2009. CC BY 2.0. https://www.flickr.com/photos/44124348109@N01/4044601053. 43

CHAPTER 3

Perception in Autonomous Driving

Abstract

In autonomous driving, the goal of perception is to sense the surrounding dynamic environment, to build reliable and detailed representation based on sensory data. In order for autonomous driving vehicles to be safe and intelligent, perception modules must be able to detect any obstacle, to recognize road surface, lane dividers, traffic signs and lights, to track moving objects in 3D, etc. Since all subsequent driving decisions, planning, and control depend on a correct perception output, its importance cannot be overstated. In this chapter, major functionalities of perception are covered, along with public datasets, problem definitions, and typical algorithms. Algorithms based on neural networks are discussed in the next chapter.

3.1 INTRODUCTION

Autonomous vehicles move in a complex and dynamic environment. To accurately and punctually perceive the surrounding physical world is critical. Sensory data from various types of sensors, including cameras, LiDAR, short-wave radar, and ultra-sonic sensors can be used. Among these sensors, cameras and LiDAR offer the most useful information. The problem of visual inference from imaging sensors is the central subject of computer vision, a sub-field of artificial intelligence. Many indispensable functionalities of perception in autonomous driving map nicely to fundamental problems in computer vision. Since the 1980s, there have been many attempts to build autonomous driving vehicles and the very first obstacle faced was perception. Significant progress has been made since then, however perception remains one of the most challenging and complex components.

3.2 DATASETS

In many fields, datasets that provide a sufficient number of samples for specific problems have proven to be an important catalyst for rapid improvement of finding solutions. They facilitate the fast iteration of algorithms based on quantitative evaluation of their performance, expose potential weaknesses, and enable a fair comparison. In computer vision, there are always datasets for individual fundamental problems, such as image classification [1, 2], semantic segmentation [1, 2, 3], optical flow [4, 5], stereo vision [6, 7], and tracking [8, 9]. These datasets are collected through crowdsourcing or synthetic approaches. They also contain different numbers of training samples each, ranging from a few to millions of labeled images. Overall, larger datasets and more realistic images lead to a less biased and more reliable evaluation of algorithm performance in practical situations.

Also, datasets have been specifically created for autonomous driving, such as KITTI [10] and Cityscapes [11]. These datasets are collected using various sensors from street scenes, depicting realistic situations facing autonomous vehicles.

KITTI datasets is a joint project between the Karlsruhe Institute of Technology (KIT) and the Toyota Technological Institute at Chicago (TTIC) in 2012. It has its own website at http://www.cvlibs.net/datasets/kitti/. The purpose of this project is to collect a realistic and challenging dataset for autonomous driving. The raw data was collected by vehicle, as shown in Figure 3.1.

Figure 3.1: KITTI car photo, adapted from the KITTI website.

The car is equipped with:

- 1 Inertial Navigation System (GPS/IMU): OXTS RT 3003,

- 1 Laser scanner: Velodyne HDL-64E,

- 2 Grayscale cameras, 1.4 Megapixels: Point Grey Flea 2 (FL2-14S3M-C), taking snapshots at a 10 Hz rate, and

- 2 Color cameras, 1.4 Megapixels: Point Grey Flea 2 (FL2-14S3C-C), taking snapshots at a 10 Hz rate.

The complete KITTI datasets consists of the following.

1. Stereo and optical flow data: A single stereo image pair is taken by two cameras at the same time. An optical flow image pair is taken by the same camera at consecutive time steps. There are 194 training image pairs and 195 testing image pairs. Approximately 50% of the pixels have ground truth displacement data. As shown in Figure 3.2, stereo data conveys depth information, while optical flow data conveys motion information.

Figure 3.2: Stereo (upper) and optical flow (lower) data [10], used with permission.

2. **Visual odometry data:** Twenty-two sequences of stereo image pairs, more than 40,000 frames, covering 39.2 km distance.

3. **Object detection and orientation data:** Manually labeled data with 3D frame notating object size and orientation. Object types include sedan, van, truck, pedestrian, cyclist, etc. (see Figure 3.3). Occlusion is present, and normally multiple objects are present in each image.

Figure 3.3: Object detection data [10], used with permission.

4. **Object tracking data:** Twenty-one training sequences and 29 testing sequences of images. The main tracking targets are pedestrian and cars.

5. **Road parsing data:** Two hundred and eighty-nine training images and 290 testing images covering various type of road surfaces: urban-unmarked, urban-marked, and urban multiple-marked lanes.

KITTI and Cityscapes datasets differ from traditionally computer vision datasets in the following rways:

- due to the use of multiple sensors and 3D scanners, high-precision 3D geometry is available, hence high-quality ground truth;

- they are collected from real world, not synthesized nor collected in a controlled lab setting; and

- they contains data for various perception tasks, such as recognizing different obstacles, in autonomous driving.

While these characteristics have made them widely popular, new algorithms are being constantly submitted and evaluated. Several new datasets of autonomous driving haven been released in recent years. These include:

- Audi Autonomous Driving Dataset (A2D2): https://www.a2d2.audi/a2d2/en.html,

- nuScenes: http://nuscenes.org,

- Berkeley DeepDrive: http://bdd-data.berkeley.edu,

- Waymo Open Dataset: http://waymo.com/open, and

- Lyft Level 5 Open Data: http://self-driving.lyft.com/level5/data.

3.3   DETECTION

Autonomous vehicles share the road with many other traffic participants such as cars, pedestrians, etc. Also on the road are obstacles, lane dividers, and myriad other objects. Fast and reliable detection of these objects (Figure 3.4) is obviously crucial. Object detection is a fundamental problem in computer vision, and many algorithms have been proposed to address it.

Figure 3.4: Car detection in KITTI, based on [34], used with permission.

The detection pipeline typically starts with a preprocessor of the input images, followed by a region of interest detector and finally a classifier that identifies the objects detected. Due to possible large variances in position, size, aspect ratio, orientation, and appearances, object detectors must, on the one hand, extract distinctive features that can separate different object classes, and on the other hand construct invariant object representation that makes detection reliable. Another import aspect of object detection in autonomous driving is speed: normally detectors must run at close to real time.

A good object detector needs to model both the appearance and the shape of objects under various conditions. In 2005, Dalal and Triggs [12] proposed an algorithm based on Histogram Of Orientation (HOG) and Support Vector Machine (SVM). The whole algorithm is shown in Figure 3.5. It passes input image through preprocessing, computes HOG features over sliding detection window, and uses a linear SVM classifier for detection. This algorithm captures object appearance by purposefully designed HOG features and depends on linear SVM to deal highly nonlinear object articulation.

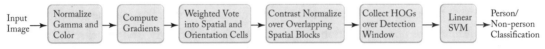

Figure 3.5: HOG+SVM. Adapted from Dalal and Triggs [12].

Articulated objects are challenging because they form a complex appearance due to their non-rigid shape. The Deformable Part Model (DPM) by Felzenszwalb et al. [13] splits objects into simpler parts so that DPM can represent non-rigid objects by composing them from easier parts. This reduces the number of training examples needed for the appearance modeling of whole objects. DPM (Figure 3.6) uses an HOG feature pyramid to build multiscale object hypotheses, spatial constellation model of part configuration constraint, and latent SVM to handle latent variables such as part position.

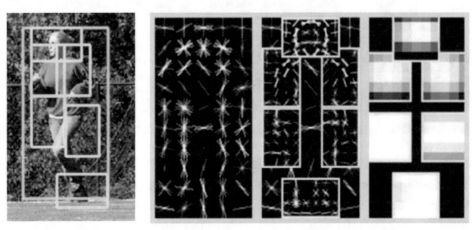

Figure 3.6: Deformable part model. Adapted from Felzenszwalb et al. [13], used with permission.

Object detection can be based on other types of sensors too, such as LiDAR [14]. Even though the performance of LiDAR-based algorthms and camera-based algorithms are similar for cars, the former have more difficulties detecting pedestrians and cyclists, probably because they are relatively small. This brings the inescapable conclusion that multiple types of sensors should be in simultaneous use and fused for better detector performance.

Indeed, autonomous vehicles must be able to navigate in traffic with pedestrians in close proximity. For obvious safety reasons, pedestrian detection is thus absolutely critical. However, human behavior is sometimes difficult to predict. Further, human appearance varies significantly and often appears with partial occlusion. There are good surveys [15, 16] available that cover various architectures. Today, most state-of-the-art pedestrian detectors use convolutional neural networks, which are discussed in the next chapter.

3.4   SEGMENTATION

Segmentation, or to be more specific, instance-level semantic segmentation, can be thought of as a natural enhancement of object detection that needs to solve sufficiently well in order for autonomous driving to be practical. Parsing image from camera into semantic meaningful segments gives the autonomous vehicle a structured understanding of its environment (Figure 3.7).

Figure 3.7: Semantic segmentation of a scene in Zurich. Courtesy of Cityscapes Dataset [11], used with permission.

Traditionally, semantic segmentation is formulated as a graph labeling problem with vertices of the graph being pixels or super-pixels. Inference algorithms on graphical models such as Conditional Random Field (CRF) are used [17, 18]. In this approach, CRFs are built with vertices representing pixels or super-pixels. Each node can take a label from a pre-defined set, conditioned on features extracted at corresponding image position. Edges between these nodes represent constraints such as spatial smoothness, label correlations, etc. (see Figure 3.8).

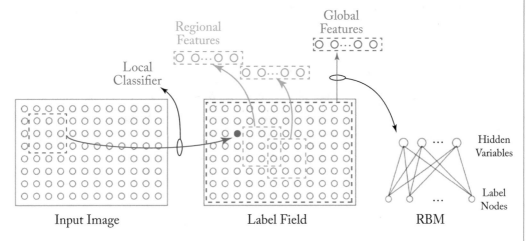

Figure 3.8: Graphical model representation of He et al. [17].

Even though CRF is a suitable approach to segmentation, it slows down when image dimension, input feature numbers, or label set size increase and has difficulty capturing long-range dependency in images, a highly efficient inference algorithm is proposed in [19] to improve speed for fully connected CRF with pairwise potentials between all pairs of pixels, other algorithms [20] aim to incorporate co-occurrence of object classes. Essentially, semantic segmentation must be able to predict dense class labels with multi-scale image features and contextual reasoning. We will discuss how deep learning approaches semantic segmentation in the next chapter.

3.5   STEREO, OPTICAL FLOW, AND SCENE FLOW

In this section, we introduce several perception techniques, including stereo vision, optical flow, and scene flow.

3.5.1   STEREO AND DEPTH

Autonomous vehicles move in a 3D world. Thus, perception that produces 3D spatial information such as depth is indispensable. Lidar generates high-precision depth data but only sparse 3D point clouds. A single image gives spatially dense information of color and texture, but not depth. Humans mostly experience 3D visual perception with two eyes. Similarly, we can gain a depth of information with a stereo camera taking pictures simultaneous at slightly different angles.

Given an image pair from stereo camera (I_l, I_r), extracting stereo information is essentially a correspondence problem where pixels in the left image I_l are matched to pixels in the right image I_r based on a cost function. The assumption is that corresponding pixels map to the same physical point, and thus have the same appearance:

$$I_l\,(p) = I_r\,(p+d),$$

where p is a location in left image and d is the disparity.

Feature-based methods replace pixel values with more distinctive features ranging from simple ones like edge and corner to sophisticated manually designed features like SIFT [21], SURF [22], etc. This leads to more reliable matching but also sparser spatial correspondence. Area-based methods exploits spatial smoothness according to the following equation:

$$d(x, y) \approx d(x + \alpha, y + \beta)$$

for fairly small (α, β). Solving for d thus becomes a minimization problem:

$$min_d\, D(p, d) = min_d \sum_{q \in N(p)} ||I_r\,(q + d) - I_l\,(q)||.$$

This can generate dense outputs with higher computation cost.

Another way to formulate the correspondence problem is optimization. Both feature-based and area-based methods are considered local since d is computed based on local information. Global methods, on the other hand, approach matching as an energy minimization problem with terms derived from a constant appearance assumption and spatial smoothness constraints. Various techniques can be used to find a global option solution, including variational methods, dynamic programming, and belief propagation.

Semi-Global Matching (SGM) [23] is one of the best-known stereo matching algorithms. It is theoretically supported [24] and also quite fast [25]. It is a global method with energy function terms calculated along several 1D lines at each pixel and also smoothness terms (see [23] for details). Recently, deep-learning-based methods have been found to have the best performance. They are discussed in the next chapter.

Once correspondence is established between stereo image pairs from two cameras of focal length f separated by a distance B (assuming the camera optical axis are aligned, which limits disparity d to simply be a scalar). A point in the image with disparity d has a depth z:

$$z = \frac{B}{d}f.$$

3.5.2   OPTICAL FLOW

Optical flow [26], as another basic computer vision problem, is defined as 2D motion of intensities between two images, which is related, but different from, the 3D motion in the physical world. It relies on the same constant appearance assumption:

$$I_t(p) = I_{t+1}(p + d)$$

but optical flow is actually more complicated than stereo. In stereo image understanding, image pairs are taken at the same time, geometry is the dominating cause of disparity, and appearance constancy is most likely to hold true. In optical flow, image pairs are taken at slightly different times, which means that motion is just one of many varying factors such as lighting, reflections, transparency, etc. Consequently, appearance constancy is likely to be violated from time to time. Another challenge for optical flow is the aperture problem (Figure 3.9): the gap between one constraint with two unknown components of d. This can be addressed by introducing smoothness constraint on the motion field of d.

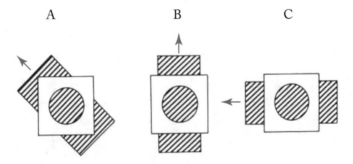

Figure 3.9: Aperture problem, adapted from https://stoomey.wordpress.com/2008/04/18/20/.

One way to alleviate potential issues caused by appearance constancy violation is replacing quadratic penalty used in [26] with robust cost function as in [27, 28].

3.5.3   SCENE FLOW

It is worth pointing out that what autonomous vehicles need is not 2D optical flow in the image plane but actual, 3D motion of objects. This is the idea behind the KITTI scene flow 2015 benchmark. Scene flow estimation is based on two consecutive stereo image pairs (Figure 3.10) where the correspondence produces not only the 3D position of points but also their 3D motion between time intervals.

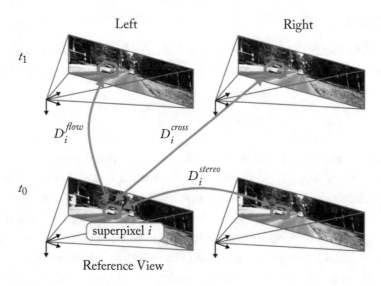

Figure 3.10: Two stereo image pairs for scene flow [29], used with permission.

To estimate scene flow, Menze and Geiger [29] use super-pixel to build a 3D parametric plane for moving objects (Figure 3.11), thereby assuming piecewise rigidity.

Figure 3.11: Estimated moving object, optical flow, and scene flow [29], used with permission.

3.6   TRACKING

The goal of tracking is to estimate object states such as location, speed, and acceleration over time. Autonomous vehicles must track various traffic participants to maintain a safe distance from them and predict their trajectories. This is especially difficult for pedestrians and cyclists because they can abruptly change direction. Tracking is generally challenging for several reasons:

- objects are often partially or fully occluded;

- objects of same class may be highly similar in appearance; and

- appearance of an object can change greatly due to pose, articulation, and lighting conditions during tracking time.

Traditionally, tracking is formulated as a sequential Bayesian filtering problem.

1. **Prediction step:** given the state of the object at the previous time step, predict the state of the object at the current time step using a motion model that describes the temporal evolution of the state of the object.

2. **Correction step:** given the predicted state of the object at the current time step, and the current observation deduced from sensor data, a posterior probability distribution of the state of the object at the current time is calculated using an observation model that represents how observations are determined by the state of the object.

3. This process goes on iteratively.

Particle filter is commonly used for tracking [30, 31]. However, the recursive nature of the Bayesian filtering formulation makes it hard to recover from a temporary detection failure. If tracking is approached in a non-iterative manner, it can be thought as minimizing a global energy function that incorporates motion smoothness constraints and appearance constancy assumptions. However, the downside of this approach is the number of object hypothesis and the number of possible trajectories per object can both be large and make finding an optimal solution computationally expensive. One way to address is using some kind of heuristic to help energy minimization [32].

Another popular formulation of object tracking is tracking-by-detection. An object detector is applied to consecutive frames and detected objects are linked across frames. These two steps both meet some uncertainty: possible missed detection and false position from detector and data association problem when resolving the combinatorial explosion of possible trajectories. These uncertainties can be naturally handled with a Markovian Decision Process (MDP). In [33], objects tracking is formulated as an MDP (Figure 3.12).

- Objects have four types of state: active, inactive, tracked, lost. $s \in S = S_{active} \cap S_{tracked} \cap S_{lost} \cap S_{inactive}$;

- ◦ When an object is detected, it is "active."

- ◦ If the detection is considered valid, the corresponding object enters the "tracked" state.

- ◦ If the detection is considered invalid, the corresponding object enters the "inactive" state.

- ◦ A "tracked" object can become "lost."

- ◦ A "lost" object can re-appear to become "tracked."

- ◦ If an object stays "lost" for sufficiently long, it becomes "inactive."

- ◦ An "inactive" object stays "inactive."

- All actions $a \in A$.

- Transition function T: $S \times A \rightarrow A$ is deterministic.

- Reward function R: $S \times A \rightarrow R$ is learned from data.

- Policy π: $S \rightarrow A$ is also learned from data.

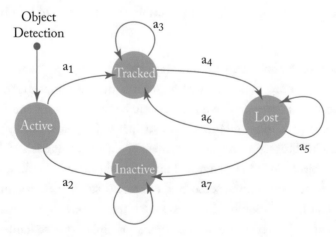

Figure 3.12: MDP formulation of tracking (based on [33]).

This algorithm runs as follows.

- In the "active" state, object candidates proposed by the detector are passed through an SVM-trained offline to decide their validity. The SVM considers features, and position of candidates and chooses action a_1 or a_2.

- In the "tracked" state, a tracking-learning-detection based algorithm uses an online appearance model to decide if object stays in the "tracked" state or goes in the "lost" state. This appearance model uses object bounding box as a template. If an object remains in the "tracked" state, its appearance model is continuously updated.

- In the "lost" state, all templates collected during an object's "tracked" states are used to decide if it goes back to the "tracked" state. If an object remains in the "lost" state for longer than a given time, it enters the "inactive" state.

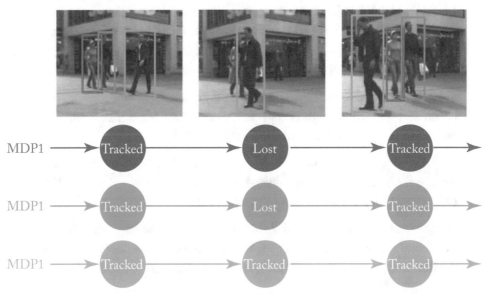

Figure 3.13: Example of MDP [33], used with permission.

This approach achieves state-of-the-art performance on KITTI dataset; see an example in Figure 3.13.

3.7   CONCLUSION

Perception as one of core components of autonomous driving relies heavily on computer vision. In this chapter, we have discussed several highly relevant areas of computer vision research and how these techniques have been applied in perception tasks. Problem definitions, datasets, common approaches, and their strength/weakness were covered to provide an appropriate understanding.

3.8   REFERENCES

[1]   Deng, J., Dong, W., Socher, R., Li, L.J., Li, K., and Fei-Fei, L. 2009, June. Imagenet: A large-scale hierarchical image database. In *IEEE Conference on Computer Vision and Pattern Recognition, 2009. CVPR 2009.* (pp. 248–255). IEEE. 51

[2]   Everingham, M., Van Gool, L., Williams, C.K., Winn, J., and Zisserman, A. 2010. The pascal visual object classes (voc) challenge. *International Journal of Computer Vision*, 88(2), pp. 303–338. DOI: 10.1007/s11263-009-0275-4. 51

[3]   Lin, T.Y., Maire, M., Belongie, S., Hays, J., Perona, P., Ramanan, D., Dollár, P., and Zitnick, C.L. 2014, September. Microsoft coco: Common objects in context. In *European Conference on Computer Vision* (pp. 740–755). Springer International Publishing. DOI: 10.1007/978-3-319-10602-1_48. 51

[4]   Baker, S., Scharstein, D., Lewis, J.P., Roth, S., Black, M.J., and Szeliski, R. 2011. A database and evaluation methodology for optical flow. *International Journal of Computer Vision*, 92(1), pp. 1–31. DOI: 10.1007/s11263-010-0390-2. 51

[5]   Janai, J., Güney, F., Wulff, J., Black, M., and Geiger, A. 2017. Slow flow: Exploiting high-speed cameras for accurate and diverse optical flow reference data. *IEEE Conference on Computer Vision and Pattern Recognition* (CVPR). 51

[6]   Scharstein, D., Szeliski, R., and Zabih, R. 2001. A taxonomy and evaluation of dense two-frame stereo correspondence algorithms. In *Stereo and Multi-Baseline Vision, 2001.* (SMBV 2001). *Proceedings. IEEE Workshop on* (pp. 131–140). IEEE. DOI: 10.1109/SMBV.2001.988771. 51

[7]   Scharstein, D., Hirschmüller, H., Kitajima, Y., Krathwohl, G., Nešić, N., Wang, X., and Westling, P. 2014, September. High-resolution stereo datasets with subpixel-accurate ground truth. In *German Conference on Pattern Recognition* (pp. 31–42). Springer International Publishing. DOI: 10.1007/978-3-319-11752-2_3. 51

[8]   Leal-Taixé, L., Milan, A., Reid, I., Roth, S., and Schindler, K. 2015. MOTChallenge 2015: Toward a benchmark for multi-target tracking. arXiv preprint arXiv:1504.01942.

[9]   Milan, A., Leal-Taixé, L., Reid, I., Roth, S., and Schindler, K. 2016. Mot16: A benchmark for multi-object tracking. arXiv preprint arXiv:1603.00831. 51

[10]   Geiger, A., Lenz, P., and Urtasun, R. 2012, June. Are we ready for autonomous driving? the kitti vision benchmark suite. In *2012 IEEE Conference on Computer Vision and Pattern Recognition (CVPR)*, (pp. 3354–3361). IEEE. DOI: 10.1109/CVPR.2012.6248074. 52

[11]   Cordts, M., Omran, M., Ramos, S., Rehfeld, T., Enzweiler, M., Benenson, R., Franke, U., Roth, S., and Schiele, B. 2016. The cityscapes dataset for semantic urban scene understanding. In *Proceedings of the IEEE Conference on Computer Vision and Pattern Recognition* (pp. 3213–3223). DOI: 10.1109/CVPR.2016.350. 52

[12]   Dalal, N. and Triggs, B. 2005, June. Histograms of oriented gradients for human detection. In *IEEE Computer Society Conference on Computer Vision and Pattern Recognition, 2005*. CVPR 2005. (Vol. 1, pp. 886–893). IEEE. DOI: 10.1109/CVPR.2005.177. 55

[13]   Felzenszwalb, P., McAllester, D., and Ramanan, D. 2008, June. A discriminatively trained, multiscale, deformable part model. In *IEEE Conference on Computer Vision and Pattern Recognition, 2008*. CVPR 2008. (pp. 1–8). IEEE. DOI: 10.1109/CVPR.2008.4587597. 55

[14]   Wang, D.Z. and Posner, I. 2015, July. Voting for voting in online point cloud object detection. In *Robotics: Science and Systems*. DOI: 10.15607/RSS.2015.XI.035. 56

[15]   Enzweiler, M. and Gavrila, D.M. 2009. Monocular pedestrian detection: Survey and experiments. *IEEE Transactions on Pattern Analysis and Machine Intelligence*, 31(12), pp. 2179–2195. DOI: 10.1109/TPAMI.2008.260. 56

[16]   Benenson, R., Omran, M., Hosang, J., and Schiele, B. 2014. Ten years of pedestrian detection, what have we learned?. arXiv preprint arXiv:1411.4304. 56

[17]   He, X., Zemel, R.S., and Carreira-Perpiñán, M.Á. 2004, June. Multiscale conditional random fields for image labeling. In *Proceedings of the 2004 IEEE Computer Society Conference on Computer Vision and Pattern Recognition, 2004. CVPR 2004*. (Vol. 2, pp. II–II). IEEE. 56

[18]   He, X., Zemel, R., and Ray, D. 2006. Learning and incorporating top-down cues in image segmentation. *Computer Vision–ECCV 2006*, pp. 338–351. DOI: 10.1007/11744023_27. 56

[19]   Krähenbühl, P. and Koltun, V. 2011. Efficient inference in fully connected crfs with gaussian edge potentials. In *Advances in Neural Information Processing Systems* (pp. 109–117). 57

[20]   Ladicky, L., Russell, C., Kohli, P., and Torr, P.H. 2010, September. Graph cut based inference with co-occurrence statistics. In *European Conference on Computer Vision* (pp. 239-253). Springer Berlin Heidelberg. DOI: 10.1007/978-3-642-15555-0_18. 57

[21]   Lowe, D.G. 1999. Object recognition from local scale-invariant features. In Computer vision, 1999. The *Proceedings of the Seventh IEEE International Conference on* (Vol. 2, pp. 1150–1157). IEEE. DOI: 10.1109/ICCV.1999.790410. 58

[22]   Bay, H., Tuytelaars, T., and Van Gool, L. 2006. Surf: Speeded up robust features. *Computer Vision–ECCV 2006*, pp. 404–417. DOI: 10.1007/11744023_32. 58

[23]   Hirschmuller, H. 2008. Stereo processing by semiglobal matching and mutual information. *IEEE Transactions on Pattern Analysis and Machine Intelligence*, 30(2), pp. 328–341. DOI: 10.1109/TPAMI.2007.1166. 58

[24]   Drory, A., Haubold, C., Avidan, S., and Hamprecht, F.A. 2014, September. Semi-global matching: a principled derivation in terms of message passing. In *German Conference on Pattern Recognition* (pp. 43–53). Springer International Publishing. DOI: 10.1007/978-3-319-11752-2_4. 58

[25]   Gehrig, S.K., Eberli, F., and Meyer, T. 2009, October. A real-time low-power stereo vision engine using semi-global matching. In *International Conference on Computer Vision Systems* (pp. 134–143). Springer Berlin Heidelberg. DOI: 10.1007/978-3-642-04667-4_14. 58

[26]   Horn, B.K. and Schunck, B.G. 1981. Determining optical flow. *Artificial Intelligence*, 17(1-3), pp. 185–203. DOI: 10.1016/0004-3702(81)90024-2. 59

[27]   Black, M.J. and Anandan, P. 1996. The robust estimation of multiple motions: Parametric and piecewise-smooth flow fields. *Computer Vision and Image Understanding*, 63(1), pp. 75–104. DOI: 10.1006/cviu.1996.0006. 59

[28]   Zach, C., Pock, T., and Bischof, H. 2007. A duality based approach for realtime TV-L 1 optical flow. *Pattern Recognition*, pp. 214–223. DOI: 10.1007/978-3-540-74936-3_22. 59

[29]   Menze, M. and Geiger, A. 2015. Object scene flow for autonomous vehicles. In *Proceedings of the IEEE Conference on Computer Vision and Pattern Recognition* (pp. 3061–3070). DOI: 10.1109/CVPR.2015.7298925. 60

[30]   Giebel, J., Gavrila, D., and Schnörr, C. 2004. A Bayesian framework for multi-cue 3d object tracking. *Computer Vision-ECCV 2004*, pp. 241–252. DOI: 10.1007/978-3-540-24673-2_20. 61

[31]   Breitenstein, M.D., Reichlin, F., Leibe, B., Koller-Meier, E., and Van Gool, L. 2011. Online multiperson tracking-by-detection from a single, uncalibrated camera. *IEEE Transactions on Pattern Analysis and Machine Intelligence*, 33(9), pp. 1820–1833. DOI: 10.1109/TPAMI.2010.232. 61

[32]   Andriyenko, A. and Schindler, K. 2011, June. Multi-target tracking by continuous energy minimization. In *2011 IEEE Conference on Computer Vision and Pattern Recognition (CVPR)*, (pp. 1265–1272). IEEE. DOI: 10.1109/CVPR.2011.5995311. 61

[33]   Xiang, Y., Alahi, A., and Savarese, S. 2015. Learning to track: Online multi-object track-ing by decision making. In *Proceedings of the IEEE International Conference on Computer Vision* (pp. 4705–4713). DOI: 10.1109/ICCV.2015.534. 61

[34]   Geiger, A., Lenz, P., Stiller, C., and Urtasun, R. 2013. Vision Meets Robotics: the KITTI Dataset. *September 2013 International Journal of Robotics Research*, 32(11), pp. 1231–1237. DOI: 10.1177/0278364913491297. 54

CHAPTER 4

Deep Learning in Autonomous Driving Perception

Abstract

In the previous chapter, we discussed perception issues in autonomous driving. In recent years, the concept of deep neural networks, also known as deep learning, has greatly changed the field of computer vision, making significant progress in solving various problems such as image classification, object detection, semantic segmentation, etc. Most state-of-the-art algorithms now apply one type of neural network that is based on convolution operation, and the field is rapidly progressing. In this chapter, we will cover selected deep-learning-based algorithms for perception in autonomous driving.

4.1 CONVOLUTIONAL NEURAL NETWORKS

Convolutional Neural Networks (CNNs), belong to one type of DNNs that uses convolution as the primary computational operator. It was reported by LeCun et al. [1] in 1988, whose inspiration can be traced back to Hubel and Wiesel's Nobel-winning work on visual cortex in 1968. They found neurons in visual cortex area V1 are orientationally selective and translationally invariant. These properties, together with the concept of local receptive fields, led to the Neo-Cognitron [2] and eventually to LeCun's LeNet [1]. CNN is a deep feedforward neural network with following properties.

- Between two layers of hidden neurons, connections are not between any two neurons in each layer, but remain "local," meaning that a neuron in the upper layer takes inputs only from neurons in the lower layer that are close, normally within a square area. This area is known as the neuron's receptive field.

- These "local" connection weights are spatially shared across neurons with the same layer. This exploits translational invariance in visual data and significantly reduces the number of parameters of the CNN model.

These properties can be thought as "implicit prior" knowledge of vision, thus making CNN quite a powerful model in solving computer vision problems. This is eloquently demonstrated in

AlexNet [3], the winner of the ImageNet image classification challenge in 2012. Since then, adoption of CNN in computer vision has accelerated and many state-of-the-art algorithms are now based on CNN. Naturally, CNN has become central to autonomous driving perception.

4.2   DETECTION

Traditionally, object detection algorithms use hand-crafted features to capture relevant information from images and a structured classifier to deal with spatial structures. This approach cannot fully exploit extremely large data volume and deal with endless variations of object appearance and shape. Girschick et al. [4] adopted a propose-then-classify approach and proved that CNN can then be used to get much better performance in object detection. Subsequent work (Fast R-CNN [5] and Faster R-CNN [6]) improved both speed and accuracy.

Faster R-CNN splits object detection into two steps that share one underlying CNN.

1. Given an input image, first generate possible regions of interest: because an object may be located in various positions, and have many scale and aspect ratio possibilities, an efficient method is needed that cuts down the number of candidates proposed, while achieving high recall rates. Faster R-CNN uses Region Proposal Network (RPN) for this purpose. RPN takes the last feature map of a CNN as input, and connects it to a hidden layer of 256-d (or 512-d) using a 3×3 sliding window and at last to two fully connected layers, one for object class, the other for object coordinates. In order to accommodate various object sizes (128×128, 256×256, 512×512) and aspect ratios (1:1, 1:2, 2:1), 3*3=9 combinations are considered at each location. For an image of size 1000 × 600, this leads to (1000/16) * (600/16) * 9~20,000 hypothesis. CNN makes this computation very efficient. Finally, we use non-maximal suppression to remove redundancy and keep about 2,000 object proposals (Figure 4.1).

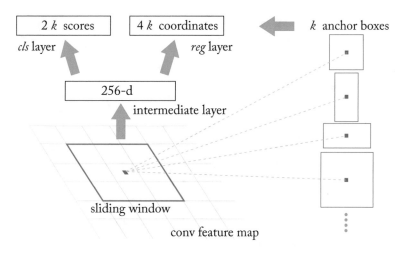

Figure 4.1: RPN [6], used with permission.

2. Given a proposed region, assess the existence of an object and its class and adjust region size, location, and aspect ratio for precision. As seen in Figure 4.2, each proposed region is first projected to a fixed-size feature map by the ROI pooling layer, then through several fully connected layers, ending as a feature vector. Finally, object class and location/size are predicted by two separate branches.

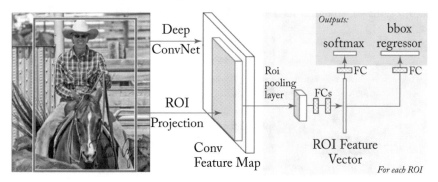

Figure 4.2: Fast R-CNN algorithm (adapted from [5]), used with permission.

There is also another group of proposal-free algorithms such as SSD [7], YOLO [8] and YOLO9000 [9]. The common theme among these algorithms is an end-to-end CNN without the proposal step. For example, SSD (Figure 4.3) uses VGG-16 network [10] as a feature extractor. By adding progressively shrinking convolutional layers on top, SSD essentially considers objects of various size and location. By predicting object position and class in one pass, SDD skips the proposal generation and the image or feature map resizing steps, thus improving performance.

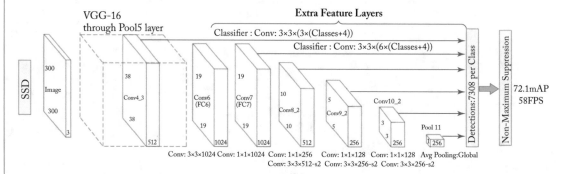

Figure 4.3: SSD network, adapted from [7].

Most proposal-less algorithms can perform object detection in real time. However, as pointed in [11], by reducing the number of proposals, Faster R-CNN can also be run in real time with similar accuracy. Proposal-based algorithms like Faster R-CNN have proven to achieve best performance on PASCAL VOC benchmarks. However, their performance dropped with KITTI. This is mainly because the KITTI dataset contains objects with a wide range of sizes as well as objects that are small or heavily occluded. To address such difficulties, Cai et al. [12] proposed a multi-scale CNN. As Figure 4.4 shows, this CNN has a "trunk" that extracts features at different scales and also "branches" that aim to detect objects of various scales.

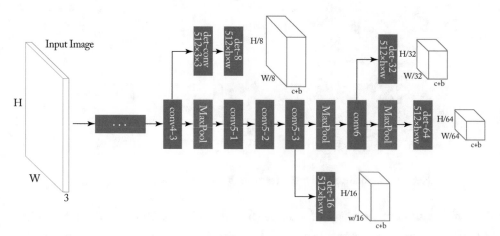

Figure 4.4: MS-CNN network, adapted from [12].

Each "branch" consists of the following key elements (Figure 4.5):

1. a deconvolution layer that increases feature map resolution for more precise localization; and

2. an ROI-pooling layer with a slightly enlarged region to capture contextual information. This can help improve classification accuracy.

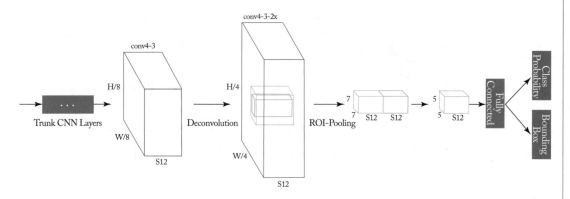

Figure 4.5: Detection branch of MS-CNN, adapted from [12].

With these adjustments, MS-CNN achieves better performance in KITTI than Faster R-CNN, especially for pedestrian and cyclist classes.

Recent improvements in detection algorithms include a point-based approach that avoids problems with pre-defined anchor boxes, or regions of interest. FCOS [20] is one such algorithm which uses a fully convolutional network and trains the classifier at each spatial location in multiple feature maps (Figure 4.6).

Without anchor boxes, FCOS can detect objects of diverse sizes and aspect ratios. It also improves recall by improving positive/negative sample balance. The complete network consists of:

- a backbone network that extracts features at increasing spatial scope and decreasing density,

- a feature pyramid network that fuses information from multiple layers, and

- shared heads between multiple feature levels that detect objects of various sizes.

Just like previous detection networks, the heads have classification and regression branches. In addition, FCOS introduces center-ness branch to suppress detection results that are off-center or of low quality. Overall, FCOS achieved state-of-art accuracy with smaller memory footprints.

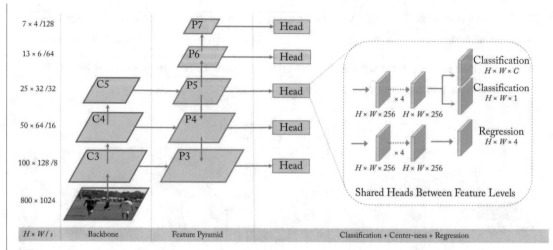

Figure 4.6: Fully Convolutional One-Stage Object Detection, adapted from [20].

4.3   SEMANTIC SEGMENTATION

In the perception module, semantic segmentation (or more broadly speaking, scene parsing) is indispensable. For example, autonomous vehicles need to understand where the road is. This requires parsing road surface out of the camera image. Semantic segmentation in computer vision, while having been studied in depth in the past, has been greatly aided by the introduction of deep learning.

Most CNN-based semantic segmentation work is based on Fully Convolutional Networks (FCN) [18]. They are designed by making the key observation that by removing the softmax layer and replacing the last fully connected layer with a 1x1 convolutional layer, CNN for image classification such as VGG-19 can be converted into an FCN. Such network cannot only accept images of any size as input, but can also attach an object/category label to each pixel.

One way to understand FCN is that it relies on a large receptive field of higher-level features to predict pixel-level labels. As a consequence, it sometimes has difficulty segmenting small objects because information from such objects are likely overwhelmed by other pixels within the same receptive field. The observation that many local ambiguities can be resolved by considering other co-occurring visual patterns in the same image shows that one key issue in semantic segmentation is a strategy to combine global image-level information with locally extracted features.

Inspired by the spatial pyramid pooling network in [19]. Zhao et al. [13] proposed a Pyramid Scene Parsing Network (PSPNet), as shown in Figure 4.7 whose main component is the pyramid pooling module is shown in the middle. The algorithm works as follows.

1. An input image is first passed through a normal CNN (PSPNet uses residual network) to extract feature maps.

2. The feature maps are passed through various pooling layers to reduce the spatial resolution down to $1 \times 1, 2 \times 2, 3 \times 3, 6 \times 6$ (this can be modified) so as to aggregate contextual information.

3. The resulting feature maps serve as context representation. They traverse the 1×1 convolution layer to shrink the feature vector size so that it can be proportional to receptive field size of feature.

4. Finally, all these feature maps for context representation are up-sampled back to the original image size and concatenated with the original feature maps out of CNN. One final convolution layer uses it to label each pixel.

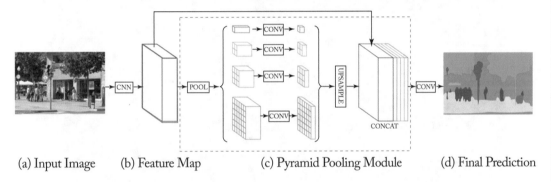

(a) Input Image   (b) Feature Map   (c) Pyramid Pooling Module   (d) Final Prediction

Figure 4.7: PSPNet (based on [13]).

PSPNet [13] has shown several experimental network design choices:

• Average pool or max pooling: experiments show average pooling is consistently better.

• Pyramid pooling module with multiple levels are consistently better than global only pooling.

• Dimensional reduction after pooling is shown to be useful.

• Training with auxiliary loss helps with optimization process of deep network.

Pyramid pooling module, together with these improvements, help make PSPNet one of the best semantic segmentation algorithms. It won 1st place in the ImageNet Scene Parsing Challenge 2016 and also has one of the best results in PASCAL VOC 2012 and Cityscapes. Some examples on Cityscapes are shown in Figure 4.8.

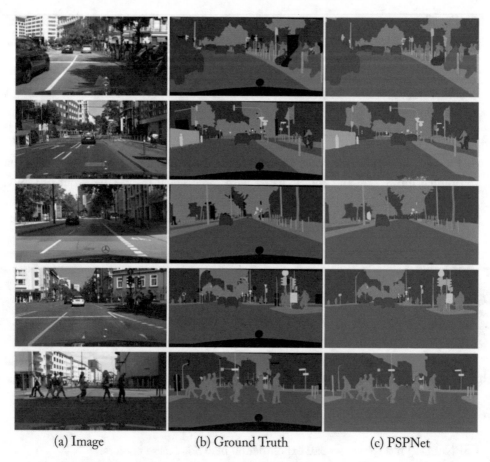

(a) Image   (b) Ground Truth   (c) PSPNet

Figure 4.8: Result examples of PSPNet, adapted from [13].

4.4   STEREO AND OPTICAL FLOW

In this section we introduce deep learning techniques applied for stereo vision and optical flow tasks.

4.4.1   STEREO

Stereo and optical flow both need to solve the correspondence problem between two input images. One simple and effective way to apply CNN in matching is the Siamese architecture such as Content-CNN proposed by Luo [14]. Content-CNN consists of two branches of convolutions layers side by side that share weights, one for the left image input, the other for the right image. Their output is merged in an inner-product layer (see Figure 4.9).

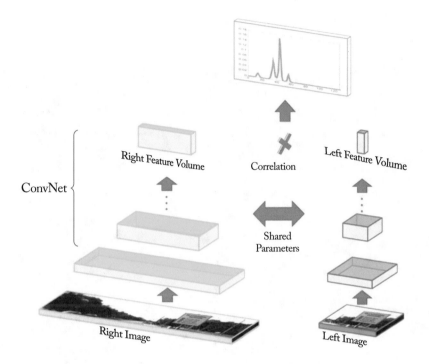

Figure 4.9: Content-CNN, adapted from [14], used with permission.

Estimation of the disparity vector at each pixel is formulated as a classification problem of 128 or 256 possible values $y \in Y$. When fed with an image pair with a known disparity y_{gt}, the network parameter w is learned by minimizing cross-entropy:

$$\min_{w} \{- \sum_{i,y(i)} P[y_{gt}(i)] \log P[y(i),w] \},$$

where

- i is the index of pixel;

- $y(i)$ is the disparity at pixel i;

- $P(y_{gt})$ is a smoothed distribution centered at y_{gt} so that the estimation error is not 0; and

- $P[y(i),w]$ is predicted probability of disparity at pixel i.

This method achieves sub-second speed on the KITTI's Stereo 2012 dataset with good estimate precision. Further post-processing can be added to enforce spatial smoothness of motion. With local windowed smoothing, semi-global block matching, and other techniques, the estimation error is reduced by 50% approximately. Such accurate 2D disparity field leads to good 3D depth estimates, as shown in Figure 4.10.

Figure 4.10: Stereo estimate on KITTI 2012 test set, adapted from [14].

4.4.2 OPTICAL FLOW

To apply deep learning in an end-to-end model of optical flow, we need to implement feature extraction, local matching, and global optimization with convolution layers. FlowNet [15] achieves this with an encoder-decoder architecture (Figure 4.11) which first "shrinks" then "expands" the convolution layers.

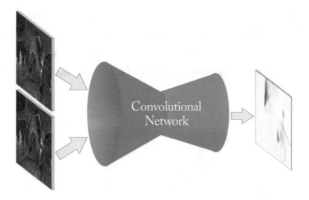

Figure 4.11: Encoder-decoder architecture of FlowNet, adapted from [15], used with permission.

There are two choices of network structures (Figure 4.12).

1. **FlowNetSimple:** This structure stacks up two images as input, and passes it through a sequence of convolution layers. It is simple, but computationally demanding.

2. **FlowNetCorr:** This structure extracts features from two images separately, then merges their feature maps together with a correlation layer, followed by convolutional layers. This correlation layer essentially computes convolution between features from two input images.

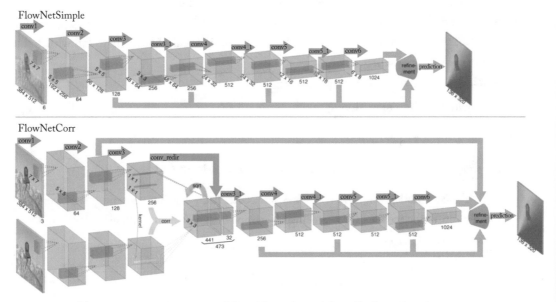

Figure 4.12: Two network architecture of FlowNet, adapted from [15], used with permission.

The "shrinking" part of FlowNet not only reduces the amount of computation, but also facilitates the spatial fusion of contextual information. However, this also lowers the output resolution. FlowNet prevents this by "up convolution" in "expanding" layers, using both feature maps from the previous layer and the corresponding layer of the same size from "shrinking" part of FlowNet (see Figure 4.13).

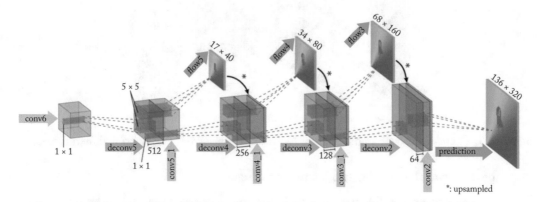

Figure 4.13: "Up convolution" in FlowNet, adapted from [15], used with permission.

FlowNet achieves competitive results on KITTI dataset, with 0.15-sec GPU run time. FlowNet 2.0 [16] further improves estimation accuracy.

Another CNN-based algorithm SpyNet [17] for optical flow takes a coarse-to-fine approach in motion estimation and uses a spatial pyramid to accomplish this. At each pyramid level, one image is warped according to current flow estimate, then an update to the flow is computed. This process iterates until full resolution flow estimate is obtained. It is argued through warping in this coarse-to-fine manner, the flow update at each pyramid level is small, thus likely to fall within the scope of layer's convolution kernel.

Assuming the following notations:

- the down-sampling operation d;

- the up-sampling operation u;

- the warping operation $w(I,V)$ of image I with flow field V; and

- a set of CNN models $\{G_0, \ldots, G_K\}$ for K levels. Each G_K has five convolutional layers and computes the residual flow v_k using up-sampled flow V_{k-1} from previous level, and resize images (I_k^1, I_k^2):

$$v_k = G_K (I_k^1, w(I_k^2, u(V_{k-1})), u(V_{k-1}))$$

$$V_k = u(V_{k-1}) + v_k.$$

During training, because of the dependency on V_k between consecutive levels, $\{G_0, \dots, G_K\}$ have to be trained sequentially one by one. During inference, we start with down-sampled images (I_0^1, I_0^2), an initial flow estimate that is 0 everywhere, and computes elements in sequence $(V_0, V_1 \dots, V_K)$ one at a time (Figure 4.14). At each level, input resized image pair and up-sampled 2-channel flow are stacked together to form an 8-channel input to G_k.

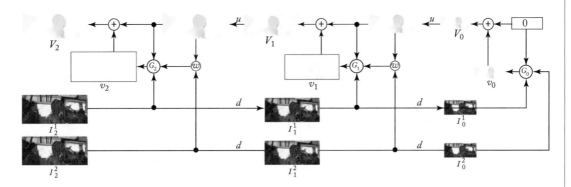

Figure 4.14: Pyramid structure of SpyNet, adapted from [17], used with permission.

SpyNet reaches state-of-the-art performance on the KITTI dataset. Most impressive is its small model size that makes it a good fit for mobile and embedded environments.

4.4.3   UNSUPERVISED LEARNING FOR DENSE CORRESPONDENCE

Both stereo depth estimation and optical flow algorithms normally require ground truth of dense correspondence for training, but such data is both expensive and difficult to collect. Therefore, in recent years, unsupervised learning has been adopted to bypass this issue. New algorithms that can learn from consecutive frames of video data have been proposed, one example is monoDepth [21] and its successor monoDepth2 [22]. Here we focus on mono-depth: its main idea is that given two images of left-view and right-view, we can define the loss function to consist of three parts.

- **Appearance matching loss:** assume corresponding pixels from these two images are similar in appearance.

- **Disparity smoothness loss:** assume disparity parameters are locally smooth with occasional discontinuities.

- **Left–right disparity consistency loss:** assume left-view disparity and right-view disparity are consistent with each other.

All of them can be computed directly from images, and thus no disparity ground truth is required.

The complete network has an "encoder" part with increasing strides, and a "decoder" part with increasing resolution. With these losses in mind, subsequent loss modules are inserted into several stages of the "decoder" part, and losses are summed together from them.

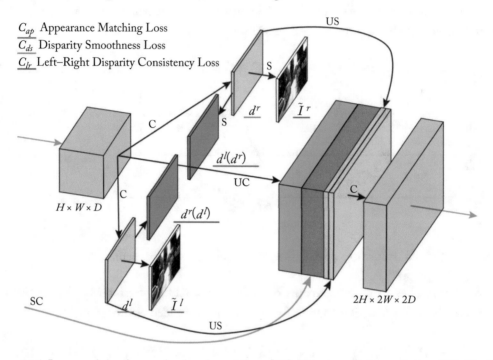

Figure 4.15: Loss module of monodepth, adapted from [21]. Notation: C: Convolution, UC: Up-Convolution, S: Bilinear Sampling, US: Up-Sampling, SC: Skip Connection.

The monodepth algorithms show better performance than earlier algorithms (Figure 4.16), and can be improved by increasing training dataset size.

Figure 4.16: **Example of monodepth, adapted from [21].**

4.5   CONCLUSION

Deep learning, as a powerful and generic model in computer vision, has made tremendous progress, and has naturally found great applications in autonomous driving. In this chapter, a subset of possible deep learning algorithms have been discussed, one or two for each perception function. Currently, these CNNs are designed and trained for specific purposes. Such homogenous building blocks of perception module may lead to a unified model architecture that is capable of performing multiple perception tasks at the same time, not only reducing redundant computation but also improving overall accuracy and robustness.

4.6   REFERENCES

[1]   LeCun, Y., Bottou, L., Bengio, Y., and Haffner, P. 1998. Gradient-based learning applied to document recognition. *Proceedings of the IEEE*, 86(11), pp. 2278–2324. DOI: 10.1109/5.726791. 69

[2]   Fukushima, K. 1988. Neocognitron: A hierarchical neural network capable of visual pattern recognition. *Neural Networks*, 1(2), pp. 119–130. DOI: 10.1016/0893-6080(88)90014-7. 69

[3]   Krizhevsky, A., Sutskever, I., and Hinton, G.E. 2012. Imagenet classification with deep convolutional neural networks. In *Advances in Neural Information Processing Systems* (pp. 1097–1105). 70

[4]   Girshick, R., Donahue, J., Darrell, T., and Malik, J. 2014. Rich feature hierarchies for accurate object detection and semantic segmentation. In *Proceedings of the IEEE Conference on Computer Vision and Pattern Recognition* (pp. 580–587). DOI: 10.1109/CVPR.2014.81. 70

[5]   Girshick, R. 2015. Fast r-cnn. In *Proceedings of the IEEE International Conference on Computer Vision* (pp. 1440–1448). DOI: 10.1109/ICCV.2015.169. 70, 71

[6]   Ren, S., He, K., Girshick, R., and Sun, J. 2015. Faster r-cnn: Toward real-time object detection with region proposal networks. In *Advances in Neural Information Processing Systems* (pp. 91–99). 70

[7]   Liu, W., Anguelov, D., Erhan, D., Szegedy, C., Reed, S., Fu, C.Y., and Berg, A.C. 2016, October. SSD: Single shot multibox detector. In *European Conference on Computer Vision* (pp. 21–37). Springer International Publishing. DOI: 10.1007/978-3-319-46448-0_2. 71, 72

[8]   Redmon, J., Divvala, S., Girshick, R., and Farhadi, A. 2016. You only look once: Unified, real-time object detection. In *Proceedings of the IEEE Conference on Computer Vision and Pattern Recognition* (pp. 779–788). DOI: 10.1109/CVPR.2016.91. 71

[9]   Redmon, J. and Farhadi, A. 2016. *YOLO9000: Better, Faster, Stronger*. arXiv preprint arXiv:1612.08242. 71

[10]   Simonyan, K. and Zisserman, A. 2014. *Very Deep Convolutional Networks for Large-Scale Image Recognition*. arXiv preprint arXiv:1409.1556. 71

[11]   Huang, J., Rathod, V., Sun, C., Zhu, M., Korattikara, A., Fathi, A., Fischer, I., Wojna, Z., Song, Y., Guadarrama, S., and Murphy, K. 2016. *Speed/Accuracy Trade-offs for Modern Convolutional Object Detectors*. arXiv preprint arXiv:1611.10012. 72

[12]   Cai, Z., Fan, Q., Feris, R.S., and Vasconcelos, N. 2016, October. A unified multi-scale deep convolutional neural network for fast object detection. In *European Conference on Computer Vision* (pp. 354–370). Springer International Publishing. DOI: 10.1007/978-3-319-46493-0_22. 72, 73

[13]   Zhao, H., Shi, J., Qi, X., Wang, X., and Jia, J. 2016. *Pyramid Scene Parsing Network*. arXiv preprint arXiv:1612.01105. 74, 75, 76

[14]   Luo, W., Schwing, A.G., and Urtasun, R.,2016. Efficient deep learning for stereo matching. In *Proceedings of the IEEE Conference on Computer Vision and Pattern Recognition* (pp. 5695–5703). DOI: 10.1109/CVPR.2016.614. 76, 77

[15]   Fischer, P., Dosovitskiy, A., Ilg, E., Häusser, P., Hazırbaş, C., Golkov, V., van der Smagt, P., Cremers, D., and Brox, T. 2015. *Flownet: Learning Optical Flow with Convolutional Networks*. arXiv preprint arXiv:1504.06852. 78, 79

[16]   Ilg, E., Mayer, N., Saikia, T., Keuper, M., Dosovitskiy, A., and Brox, T. 2016. *Flownet 2.0: Evolution of Optical Flow Estimation with Deep Networks*. arXiv preprint arXiv:1612.01925. 80

[17]   Ranjan, A. and Black, M.J. 2016. *Optical Flow Estimation using a Spatial Pyramid Network*. arXiv preprint arXiv:1611.00850. 80, 81

[18]   Long, J., Shelhamer, E., and Darrell, T.,2015. Fully convolutional networks for semantic segmentation. In *Proceedings of the IEEE Conference on Computer Vision and Pattern Recognition* (pp. 3431–3440). DOI: 10.1109/CVPR.2015.7298965. 74

[19]   He, K., Zhang, X., Ren, S., and Sun, J. 2014, September. Spatial pyramid pooling in deep convolutional networks for visual recognition. In *European Conference on Computer Vision* (pp. 346-361). Springer International Publishing. DOI: 10.1007/978-3-319-10578-9_23. 74

[20]   Tian, Z., Shen, C., Chen, H., and He, T. 2019. FCOS: Fully convolutional one-stage object detection. In *ICCV 2019*. 73, 74

[21]   Godard, C., Aodha, O. M., and Brostow, G. J. 2017. Unsupervised Monocular Depth Estimation with Left-Right Consistency. *IEEE Conference on Computer Vision and Pattern Recognition (CVPR)*, Honolulu, HI, pp. 6602–6611. DOI: 10.1109/CVPR.2017.699. 81, 82, 83

[22]   Godard, C., Mac Aodha, O., Firman, M., and Brostow, G. J. 2019. Digging into self-supervised monocular depth estimation. In *Proceedings of the IEEE International Conference on Computer Vision, 2019,* pp. 3828–3838. DOI: 10.1109/ICCV.2019.00393. 81

CHAPTER 5

Prediction and Routing

Abstract

*In this chapter, we will describe the **Prediction** and **Routing** modules with an emphasis on how they are integrated in the autonomous vehicle planning and control framework. The Prediction module is responsible for predicting the future behavior of surrounding objects identified by the Perception module. It produces predicted trajectories that are fed into the downstream planning and control modules. The Routing module we describe here is a lane level routing based on lane segmentation of the HD-maps. Routing simply informs the autonomous vehicle how to reach its destination by following a sequence of lanes on the HD-maps. Its output is also sent to the downstream planning and control modules.*

5.1 PLANNING AND CONTROL OVERVIEW

In this subsection, we provide an overview of the planning and control architecture as well as its submodules.

5.1.1 ARCHITECTURE: PLANNING AND CONTROL IN A BROADER SENSE

As shown in Figure 5.1, the *Map and Localization* module consumes raw sensor data such as point cloud and GPS. It then converts them into knowledge as to the location of the autonomous vehicle. The *Perception* module is responsible for detecting objects in the vicinity of the autonomous vehicle. These two modules are focused on perceiving the objective world, whereas the other modules, including *routing*, traffic *prediction*, behavioral *decision*, motion *planning*, and feedback *control* are involved in the subjective perspective as to how the autonomous vehicle predicts the behavior of the external environment and how the autonomous vehicle plans to move.

Modules in Figure 5.1 share a central clock. Within a clock cycle (also called a "frame"), each module independently fetches the most recently published data from its upstream modules, performs its own computation and then publishes the result for downstream modules to consume.

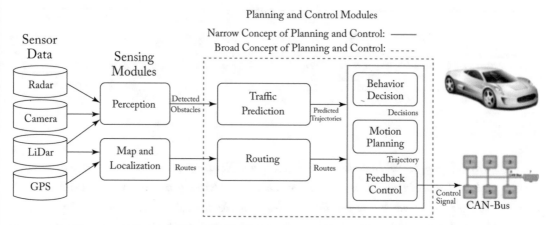

Figure 5.1: Planning and control modules under narrow and broad concepts.

As part of a complex system which involves both hardware and software, the autonomous driving software depends on the cooperation of different modules, including computing hardware, sensor integration, perception, traffic prediction, as well as motion planning and control to ensure safety and reliability. The cooperation of these modules, especially that between the perception module and the planning and control modules, is of critical importance.

A successful collaboration among these modules is can be achieved by effectively minimizing the scope of each module and the extent of the problem each module is designed to solve. In the next sub-section, we will examine how we define these functional modules in a divide-and-conquer fashion following the natural data flow and the gradual concretization of the problem. Under this approach, prediction and routing modules both belong to the broader category of planning and control while serving as data dependencies modules, which we describe in detail later in this chapter. Traditional planning and control modules, which include behavioral decision (*Decision*), motion planning (*Planning*), and feedback control (*Control*), will be discussed in the Chapter 6.

While the techniques described in this chapter and the next have been widely adopted in the design of autonomous driving systems, end-to-end solutions are becoming increasingly possible, not to say mandatory, due to the rise of Artificial Intelligence technologies. We will present these state-of-art end-to-end solutions in Chapter 7 of this book. Our emphasis will be placed on presenting a coherent and complete working solution to the broader problem of planning and control, meaning, starting from the various heterogeneous sensor inputs and the map, in other words the vehicle's objective consciousness of the surrounding world, to compute the actual manipulation of the vehicle, along with all the intermediate decisions.

5.1.2   SCOPE OF EACH MODULE: SOLVE THE PROBLEM WITH MODULES

Let us examine the function of several modules in Figure 5.1 in turn. The *routing module* issues top-level navigation commands. Routing can be understood as navigating from the originating position to the destination position by explicitly following a series of lanes. However, even though it resembles the function performed by traditional navigational map services such as *Google Maps*, the routing module in autonomous vehicle must entail more details and usually depends on HD maps customized for autonomous driving, therefore making it significantly different and more complex than traditional navigational mapping services.

Next, we turn our attention to the *traffic prediction module* (also referred to as the *prediction module*). The input to the prediction module comprise the perceived objects, along with their "objective" attributes like position, velocity, and category (e.g., vehicle, cyclist, or pedestrian). The prediction module then computes the predicted trajectories for each perceived object and passes them along to the behavioral decision module.

A predicted trajectory includes both spatial and temporal information, which will be used by the downstream modules. In previous works [1], the prediction function has been implemented as a peripheral software library either in the perception module to fine tune its output, or in the decision/planning modules to preprocess the detected input objects. The merit of implementing prediction as a software library is that software libraries do not need to periodically consume data from upstream and publish data to downstream and they do not have states or memories. The prediction software library exposes some API's for function call and returns the computation result. Hence, a software library is usually less complex in terms of computation logic. A module in the autonomous driving system, on the other hand, has to consume upstream results and publish its computation results in a periodic, frame-by-frame fashion. Modules usually need to maintain states that are memories of previous frames to enhance the accuracy of the computation.

As autonomous driving techniques evolve and autonomous vehicles hit the road in a more realistic sense, the industry has had to take notice of the importance of traffic prediction, and hence in state-of-the-art systems, the traffic prediction module is mostly implemented as an actual module rather than as a software library [4, 5, 6, 7]. In this chapter, we take a modern machine learning-based approach to formulate and solve the traffic prediction problem.

The module immediately downstream from the traffic prediction is the *behavior decision module*, which serves as the "co-pilot" in our autonomous driving system. It takes inputs from both the traffic prediction and the routing modules. From these inputs, the behavioral decision module generates commands, which determines how the vehicle should be controlled. Possible commands include *follow the vehicle in front on the current lane*, *stop behind a traffic light stop line and wait for the pedestrian to finish crossing*, or *yield to a cross vehicle at a stop sign*.

The behavioral decisions include decisions for the autonomous vehicle itself as well as behavioral assements of any perceived or map object. More specifically, assume, for example, that another vehicle has been detected in the same lane as the autonomous vehicle. The routing module would command the autonomous vehicle to remain in the current lane. The decision of the autonomous vehicle itself (a.k.a., *synthetic decision*) could be remain in lane, while the decision of the perceived vehicle in front (a.k.a., *individual decision*) could be follow the vehicle in front. The behavioral decision for each individually perceived obstacle will be converted to optimization constraints and costs in motion planning.

The behavioral decision for the autonomous vehicle itself is synthesized from all these individual behavioral decisions, and is thus referred to as *synthetic* decision. Such a synthetic main decision is necessary for determining the end state motion conditions in motion planning. The detailed design of the behavioral decision output command set could be different depending on the implementation. Modern autonomous vehicle systems are more inclined to designing and implementing the logic of behavioral decision as an individual module. However, in fact, in approaches, the logic and role of behavioral decision are incorporated into downstream modules such as motion planning [1, 2, 3].

As we mentioned the importance of collaboration of the modules, the logics of the upstream behavioral decision module and the downstream motion planning module should be coherent. This means that the motion planning module should precisely honor the behavioral decisions proferred by the behavioral decision output and integrate them accurately when making trajectorial plans for the autonomous vehicle. While the command set of behavioral decisions is meant to cover as many traffic behavioral scenarios as possible, it is not necessarily *complete*. Naturally, there are certain vague scenarios where even human drivers would not have an explicit behavior decision but rather a vague sense of collision avoidance. The explicit command set of behavioral decision is beneficial for diagnosis and debugging, but what actually matters is how these behavioral decisions are transformed into certain constraints or costs in motion planning. In the worst or oddest scenario where a reasonable individual decision cannot be made, the implicit cost of collision avoidance would have to be the *default* individual decision to pass down to the motion planning module.

Simply speaking, the motion planning module provides a planned path or trajectory to move from point A to point B. It is an optimization problem to search for a local path from point A to point B, where point A is usually the current location and point B could be any point in a desired local region, for example, any point that sits on the desired lane sequences. Motion planning takes the behavioral decision output as *constraints* and the routing output as *goals*.

Compared with *behavioral decision*, the problem that *motion planning* solves is more concrete. It must compute the trajectory along with trajectory points which consists of location, heading, velocity, acceleration, curvature, and even higher-order derivatives of these attributes. As we emphasize the collaboration among these modules, the motion planning module must enforce two important rules. First, the planned trajectory has to be consistent among consecutive planning cycles, ensuring that the trajectories of two consecutive cycles do not dramatically change if the external factors have not changed much. Second, motion planning has to ensure that the planned trajectory is executable by the downstream feedback control module, which usually indicates that attributes like curvature or its derivative have to be continuous and sufficiently smooth that they do not violate any physical control limits.

Note that, in Figure 5.1, mapping and localization info, as well as perception output, are fed directly into the motion planning and behavioral decision modules. While this could be seen as redundant in system design, it helps to ensure security as a backup for the traffic prediction. Also, during the processing of traffic prediction, new obstacles might be detected. In both cases of prediction failure and newly introduced obstacles while prediction is computing, the redundant perception information, along with easily accessible mapping and localization utility library, will help ensure that the behavioral decision and motion planning modules will at least have some basic object information to initiate the necessary steps for collision avoidance.

In the backend is the feedback control module which directly communicates with the vehicle control via the controller area network (CAN bus). Its core task is to consume the trajectory points of the planned trajectory and computes the actual drive-by-wire signals to drive the brakes, the steering wheel, and the gas. Such computation is usually performed so that the actual vehicle path conforms to the planned trajectory as closely as possible, while also keeping in consideration the physical model of the vehicle and of the road.

The modules described above are the core modules within the general concept of autonomous vehicle planning and control. The philosophy behind this partitioning [1, 2] is to effectively and reasonably decompose the complex problem of autonomous driving planning and control into a series of sub-problems. When each module is focused on solving its own problem, the complexity of the software development can be greatly reduced by modularization and parallelization. At the same time, the efficiency of research and development is hence significantly improved. This is the hallmark of our proposed solution. In essence, behavioral decision, motion planning, and feedback control are solving the same problem at different levels. Given their positions along the data stream flow, their computation results are relying on each other. An important consideration when implementing these modules is to keep the computation consistent and coherent. A general rule of thumb is: when a conflict happens, it is best to push the upstream module to solve the conflict, rather than make the downstream module adapt.

In the following sections, we will describe in more detail the modules from upstream to downstream (left to right in Figure 5.1). Again, we will emphasize the problem definition and formalization under specific scenarios that each module will face, rather than enumerate all possible solutions. We will provide one or two viable solutions by presenting them formally.

With this unique approach, we aim to bring a holistic picture depicting a comprehensive solution to the general planning and control problem in autonomous driving.

5.2   TRAFFIC PREDICTION

As the direct upstream module of planning and control modules, the *traffic prediction module* (*Prediction*) aims to predict the behavior of detected objects in the near future, computing the details of the prediction with spatial-temporal trajectory points, and passing them along to downstream modules.

Usually the detected perception obstacles have attributes of position, velocity, heading, acceleration, etc. Considering simple physical rules along with attributes, an immediate prediction could be reasonably made. However, the objective of traffic prediction is not simply the immediate prediction given the physical attributes, but more a behavioral level prediction which usually spans for a period of a few seconds. Such predictions have to take multiple factors into consideration, such as historical behavior, surrounding scenarios, and map features. For example, at a traffic junction as shown in Figure 5.2, traffic prediction needs to determine whether the vehicle will go straight through the junction or make a right turn, and whether the pedestrian on the curb will cross the junction or remain still. These behavioral predictions could be formalized into classification problems and solved by machine learning methods [8, 9, 10]. However, mere behavioral level prediction is not sufficient since the actual outputs of traffic prediction are predicted trajectories which consists of trajectory points with timing information, speed, and headings. Therefore, we further formalize the traffic prediction problem into two sub-problems.

- **Classification problem for categorical road object behaviors:** For example, will a vehicle change lane or remain in its current lane, will a pedestrian cross an intersection, etc.

- **Regression problem for generating the predicted path with speed and time info:** For example, when crossing intersection, the speed of a pedestrian might not change much, but when a vehicle makes a turn, it usually decelerates before accelerating, at rates which depend on the length and curvature of the turn.

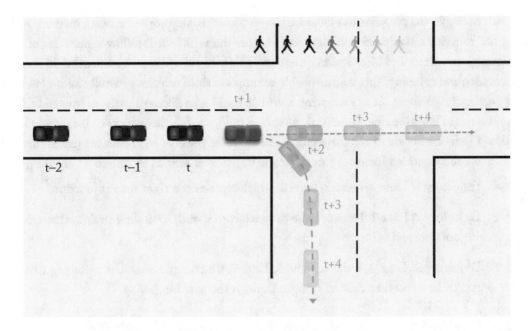

Figure 5.2: Traffic prediction for objects on the road.

5.2.1   BEHAVIOR PREDICTION AS CLASSIFICATION

The behavioral prediction problem for road objects is usually dependent on the type of the object. A vehicle's behavior might be *remain in current lane*, *make a turn*, or *change lane*, while cyclists and pedestrians have significantly wider ranges of behaviors. Given the distinctions between these behaviors, the machine learning based model must normally be customized for each category instead of applying a unified model. We will focus on the behavioral prediction for vehicles since the exclusion of pedestrians and bicycles vehicle behaviors are more predictable.

Actually, even behaviorally predicting vehicles is not a trivial problem. At first glance, one could build a classification model on three simple actions: remain in lane, switch lanes, or make a simple turn. However, our experience shows that this approach is not scalable because the real world has more complexities: for example, there could be multiple right or left turn lanes, and intersections are not always four-way. In addition, there are scenarios where the current lane naturally yields into a right turn: there are no other choices without violating traffic rules. Therefore, we cannot choose the categories of behavior based on distinctive maps or scenarios since it would make the classification categories (called "labels" in the classification problem) overwhelmingly complex and not scalable.

To decouple the classification label with map scenarios, we propose a novel method of defining the behavior classification problem as "whether the vehicle will follow a finite set of lane sequences given current and historical information." This method will somehow depend on the lane segmentation and mapping. This is a reasonable assumption since vehicles generally follow lanes on a map as they move along. At any moment, a vehicle could take different paths as it would follow different series of lanes (or lane sequences). As shown in Figure 5.3, the vehicle is currently in Lane 1 at time t (left- hand side of the diagram). There are three possible legal lane sequences that the vehicle could follow, which indicates three possible trajectories defined by different behaviors.

1. **Trajectory 1:** Lane 1, Lane 2, Lane 3, which represents a right turn at junction;

2. **Trajectory 2:** Lane 1, Lane 6, Lane 8, which corresponds to driving straight through the junction; and

3. **Trajectory 3:** Lane 1, Lane 4, Lane 5, Lane 7, which represents first switching to a parallel lane and then driving straight through the junction.

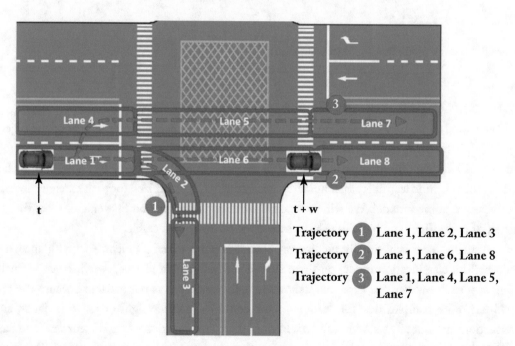

Figure 5.3: Formulating a behavioral traffic prediction problem as a binary classification problem with regard to lane sequences.

Assume that at time *t+w*, the object vehicle is now at Lane 8, through the sequence (Lane 1, Lane 6, and Lane 8). Then Trajectory 2 has a positive label, while the other two trajectories, Trajectory 1 and 3, bear a negative label. With the benefit of this innovative approach of "classifying whether a certain lane sequence will be taken," we can extract trajectories and assign them with a positive or a negative label by replaying the data (which is usually recorded as an ROS bag file). One might also wonder the following: if at time $t+w$, the vehicle is located at the intersection of Lane 2 and Lane 6, then what are the labels for Trajectory 2 and 3? It turns out that under these circumstances, both trajectories could be labeled as positive since there is simply not enough information to distinguish whether the vehicle will process straight or take a right turn. The model will likely output two predicted trajectories.

To conclude, we can formulate the behavioral prediction problem of on-road vehicles into a binary classification problem of "whether the vehicle will take a certain lane sequence." We can then build one unified machine learning model (a binary classification model) with this notion. The only dependency we rely on is the definition and segmentation of lanes, which is inevitable since the vehicle mostly follows lanes and so do our predicted trajectories.

Here we impose an assumption that vehicles follow certain logical or structural sequence of lanes. This assumption might seem quite restrictive at first glance. However, it is very important for a machine-learning approach to start with learning certain structured "reasonable" behaviors rather than learning from unrestricted behaviors. In addition, since we are adopting a learning approach, the amount of data we could accumulate places a restriction on the complexity of the learning model we can select. Therefore, for traffic prediction in autonomous driving, it is actually easier to start with learning and understanding these logic and legal behaviors first. Indeed, "illegal" behaviors (such as following logically unconnected lanes) could happen. However, once we have a good model of the legal behaviors and enough training data, we could ease the learning model to include these aberrant behaviors. For example, one could lift the restriction on lane sequence, but only place constraints on the starting and ending lane. In all, building a traffic prediction model based on lane sequence is a very effective approach for most cases.

Feature Design for Vehicle Behavior Prediction

As we discussed, the labeling of the lane-sequence-based classification problem, feature engineering, the process of using domain knowledge to extract features from raw data, is another critical aspect of building a working machine learning classifier. With the vehicle behavioral prediction problem, we could consider the following three categories of possible features (as shown in Figure 5.4).

1. **Vehicle history features:** We could consider a historical window of w frames. For each frame, the vehicle which we want to predict will be represented by its absolute position as well as its relative position to the lane. This category of features could

represent historically how the vehicle will have moved along the current lane or even previous lanes.

2. **Lane sequence features:** Note that the expanded lane sequence is an instance which we want to classify as "will take" or "will not take." Therefore, we sample some points along this designated lane sequence. For each of these sampled "lane points," we can compute its attributes which represent the expanded lane sequence shape. For example, the heading, curvature, and distance to boundary of each lane point relative to the lane to which it belongs, could be extracted as lane sequence features. This category of features could represent the shape of the designated lane sequence to be taken.

3. **Surrounding object features:** This category of feature is harder to compute and it captures the objects surrounding the vehicle (whose behavior will have to be predicted). This is because sometimes not only the lane shape or the vehicle historical pose but also the surrounding objects, will determine its future behavior. For example, if we consider the left- and right-adjacent/parallel lanes, we could project the target vehicle to the left and right adjacent lane. Then we could compute the forward distance between them.

A detailed feature vector proposal is listed in Figure 5.4. Conceptually speaking, for example, if a vehicle has been moving closer toward the right boundary of lanes with its heading also gradually leaning toward the right, it is highly possible that the lane sequence which corresponds to a right-turn will be taken than the lane sequence which corresponds to going straight, if there are no significant surrounding objects which may deter the vehicle from doing so.

Model Selection for Vehicle Behavior Prediction

Even though the above-mentioned features cover most of the information pertaining to the vehicle and its surroundings, it is not meant to be a complete feature set, but rather a suggestion of feasible feature set based on our experience. Therefore, these features might be adapted to specific machine learning models. There are two types of models we can utilize to serve as the purpose of predicting the behavior.

1. **Memory-less models:** These include models such as SVM [11] or DNN [12]. They are memory-less because the model remains the same once trained. Further, the output is not dependent on previous input instances. With this type of models, if we want to capture historical information, we need to explicitly encode them into the features. For example, we may take multiple historical frames of the vehicle information, extract features from each of these frames, and use these extracted features

for both prediction and training. In fact, the proposed feature set in Figure 5.4 takes vehicle historical information into consideration.

2. **Memory models:** These include models such as Long Short-Term Memory (LSTM) models with RNN (Recursive Neural Networks) structures [13, 14]. These models have memories since the output will be dependent on the input. However, they are much more difficult to train. With models like RNN, the input could just be the current frame information such as current vehicle feature and current surrounding object features. The model will somehow "memorize" previous inputs with model parameters, and these previous inputs will influence the current output.

Which model to choose is dependent on the specific scenarios. When the mapping and surrounding environment are not overly complex, it might be just enough to use the memory-less models. If the traffic condition is more involved, one might want to leverage memory models such as RNN to fully handle the historical information. As for an engineering implementation, memory models are easier to implement online since they only take current information as input and the model itself memorizes historical info. Memory-less models are usually more difficult to implement in an online system. The reason is that historical information usually must be fed into the model as features, and the online system must store historical information online for feature extraction. Ideally, the time window "w" could be the longest time expected for the perception module to track an object. In a typical traffic prediction module, "w" could be set to a fixed length such as 5 sec. The predicted trajectory should cover either a minimum distance or a minimum time. If we choose 5 sec as the "memory" window, the predicted trajectory could be as high as 5 sec, but it would be more reliable to assume a shorter window, such as 3 sec. Note that the exact theoretical limit of the amount of historical data necessary for accurately predicting the future behavior is beyond the scope of this book. The important metrics in behavioral prediction are the *precision* and *recall*. *Precision* means how many of all the predicted trajectories will actually be taken by the object vehicle. *Recall* means how many of all the actual behavioral trajectories, will have been predicted. Since the traffic prediction module outputs predicted trajectories at each frame, these two metrics are computed by aggregating the predicted trajectories in all the frames.

5.2.2 VEHICLE TRAJECTORY GENERATION

Once the behavior of a vehicle has been determined, the prediction module needs to generate the actual spatial-temporal trajectory which follows the predicted lane sequence. One possible simple solution is based on physical rules and certain assumptions. We propose using a Kalman-Filter to track the lane-based map coordinates of the vehicle. The core underlying assumption is that a vehicle will gradually follow the center line (also known as "reference line") of lanes. Therefore, we use a Kalman-Filter to track the (s, l) coordinates of the vehicle's predicted points on the trajectory. Simply speaking, s represents the distance along the central reference line of a lane (the longitudinal distance), and l represents the lateral distance which is perpendicular to the s direction at any point. The lane-based map coordinate system will be expanded upon in the discussion on motion planning (Chapter 6). The motion transformation matrix of the Kalman-Filter is:

$$\begin{pmatrix} s_{t+1} \\ l_{t+1} \end{pmatrix} = A \cdot \begin{pmatrix} s_t \\ l_t \end{pmatrix} + B \cdot \begin{pmatrix} \Delta t \\ 0 \end{pmatrix}, \text{ where } A = \begin{pmatrix} 1 & 0 \\ 0 & \beta t \end{pmatrix} \text{ and } B = \begin{pmatrix} v_s & 0 \\ 0 & 0 \end{pmatrix}.$$

For each predicted lane sequence, we could maintain a Kalman-Filter to track the predicted trajectory for this specific lane sequence. In the state-transfer matrix A, how fast the predicted trajectory will approximate the central reference line is controlled by the parameter β_t. With each prediction cycle, β_t could be adjusted in the Kalman-Filter measurement update, and therefore the speed at which the vehicle approximates the central line will be affected by historical observations (measurement). Once β_t has been fixed, we can do the prediction step of the Kalman-Filer for a certain number of steps and generate the trajectory point for each future frame of the prediction trajectory.

In addition to the above-mentioned rule-based method, possible machine learning-based solutions could also be used in trajectory generation. The advantage of a machine learning-based model in trajectory generation is that it could leverage historical actual trajectories and aim to generate trajectories more like the history than the rule-based trajectories. *Regressions models* in machine learning are appropriate here in this scenario. One could feed historical vehicle information as input features and try to build models to capture the actual paths of vehicles. However, our viewpoint is that the actual trajectory is much more difficult to model and it is of less importance than the behavioral itself. We mention this possibility and interested readers can explore the related work in [18]. One might note here while drawing the actual trajectory for an objective vehicle, we take a simple reference-line based approach here. In fact, all the motion planning techniques for computing motion trajectories could be applied here for obtaining the trajectories for other object vehicles.

In conclusion, we formalized the traffic prediction problem into two phrases: first predict the behavior and then computes the actual trajectory. The first behavior prediction problem is articulated as a binary classification problem on each possible lane sequence, while the second problem of computing actual spatial-temporal trajectories could borrow from techniques in motion planning. In the behavior prediction problem, interactions among different object vehicles have not been explicitly considered, since introducing the mutual impact of various vehicles would bring an exploding effect of the complexity. However, there is one viewpoint that if the prediction frequency is high enough, interactions among object vehicles could be implicitly incorporated.

Vehicle History Features	Lane Features	Surrounding Obstacle Features
Consider [t-w+1, t] frames, and for each frame, extract the following features: ➢ Longitude and latitude position on the lane; ➢ XY-based position; ➢ Speed, heading, and acceleration; ➢ Heading and curvature of the projected lane reference point; and ➢ Relative distance to lane boundaries. In addition, vehicle's length, width, and height are also extracted as features.	Consider v points sample along the central reference longitudinal direction of the target lane sequence to be classified, extract the following features from each lane point: ➢ Relative longitudinal and lateral position; ➢ Heading and curvature; ➢ Distance to the left and right boundaries; and ➢ Lane turn type of the sampled lane point.	Consider obstacles on two specific lane sequences: the target lane sequence to be classified and the current lane sequence (i.e., natural expansion of successor lanes). These two sequences might be the same. Project the target object on each lane sequence, and consider this projection as reference. Find the closet vehicles before and after this reference position. Extract features from these two vehicles: ➢ Relative longitudinal position to the projected reference point; and ➢ Lateral position, speed, and heading of these closest front and rear vehicles.

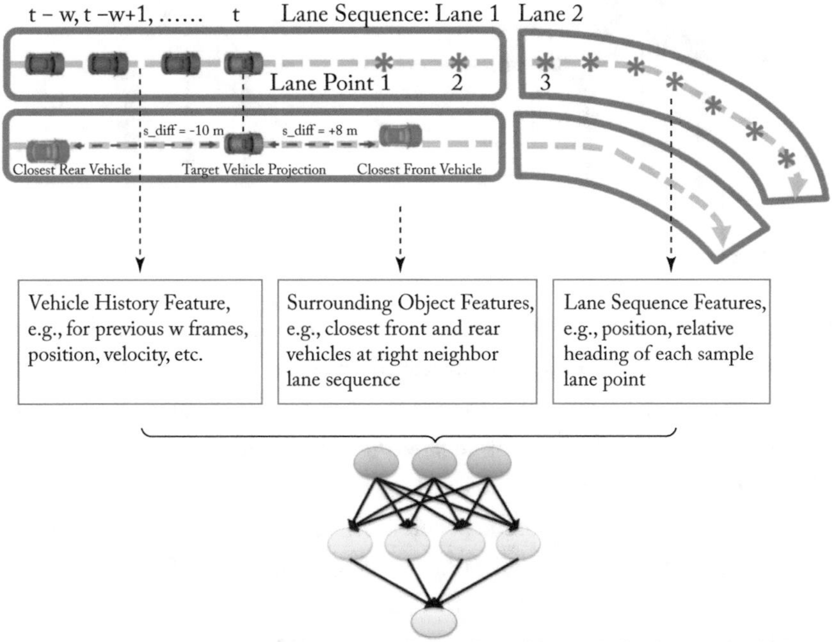

Figure 5.4: Three categories of features for classifying if a lane sequence will be taken.

5.3 LANE LEVEL ROUTING

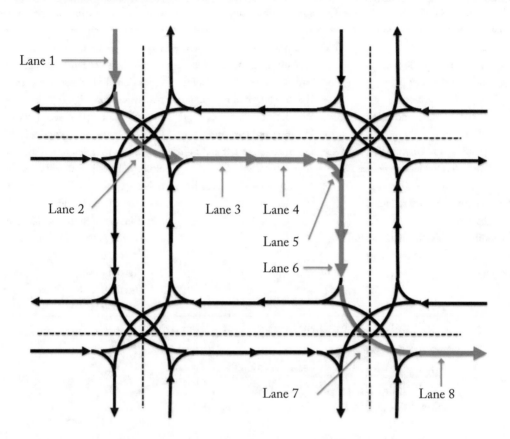

Figure 5.5: Routing output on lane levels defined by the HD map.

On top of the planning and control modules is the lane-level routing module, which we call *Routing* for short. Here the routing module is significantly different from navigational maps as Google Map. Traditional navigational maps solve the problem of getting from point A to point B via a series of roads. The minimum element for such navigation could be a specific lane on a specific road. These lanes and roads are naturally defined by the real road signs and segmentations. Even though the routing problem for autonomous vehicle also solves the problem of getting from A to B, its output is not for human drivers, but rather serving as an input of the downstream modules such as behavioral decision and motion planning. Therefore, the extent of lane-level route planning has to reach the level of lanes defined by HD-maps. These HD map-defined lanes are not the same as the naturally divided lanes or roads. As shown in Figure 5.5, the arrows represent the HD map level lane segmentation and directions. *lane*$_1$, *lane*$_2$...... *lane*$_8$ constitute a routing output sequence. One can easily see that the HD map-defined lanes are not necessarily natural lanes and correspond to actual lane markers. For example, $lane_2$, $lane_5$...$lane_7$ represent the "virtual" turning lanes as per defined by the HD map. And, similarly, a relatively long lane on a natural road could be segmented into several lanes (for example $lane_3$, $lane_4$).

As the top upstream module of the broad planning and control modules, the output of routing is heavily dependent on the creation of HD map. Given the "Road Graph" and lane segmentations defined in the HD map, and under certain predefined optimal policy, the problem that routing aims to solve is to compute an optimal lane sequence from source to destination for the autonomous vehicle to follow:

$$\{(lane_i, \text{start\_position}_i, \text{end\_position}_i)\},$$

where $(lane_i, \text{start\_position}_i, \text{end\_position}_i)$ is called a routing segment. A routing segment is identified by its belonging $lane_i$, andstart_position_i, end_position_i represent the starting and ending position along the central reference line of the lane.

5.3.1   CONSTRUCTING A WEIGHTED DIRECTED GRAPH FOR ROUTING

A distinctive characteristic for autonomous driving routing is that the routing module has to take into consideration the difficulties of certain motions for autonomous vehicles to execute while planning the optimal routes. This is a significant difference from traditional navigational map services such as Google Maps. For example, autonomous driving routing will avoid switching to parallel lanes since the motion planning module will require more space and time to fulfill this motion and it is in the best interest of safety to avoid generating routing segments which do require such short-distanced lane switching. Therefore, we shall assign a high "cost" for such possible routes. In short, the difficulties for an autonomous vehicle to perform certain actions might be very different comparing with human drivers, and hence the routing module will be customized to adapt to the driving of autonomous vehicle motion planning module. In this sense, routing output for autonomous vehicles is not necessarily the same as ordinary navigational routing outputs for human drivers.

We abstract the autonomous vehicle HD map-based routing problem, into a shortest path search problem on a weighted directed graph. The routing module will firstly sample several points on the HD map lanes within a certain proximity of the autonomous vehicle's current location. These points are called "*lane points.*" They represent possible locations on a lane the autonomous vehicle might visit while following the lane. There are directed edges connecting the lane points which are proximate to each other (see Figures 5.6 and 5.7). If we do not allow backing on a lane (a reasonable assumption), lane points are only connected to each other along the direction of the lane. The weight of the edge connecting the lane points represent the potential cost for the autonomous vehicle to move from the source lane point to the destination lane point. The sampling frequency of lane points has to ensure that even short lanes will get sufficiently sampled. Edge connections between lane points have an obvious characteristic of locality. Adjacent points along the direction of lane are naturally connected with a directed edge with the same direction as the lane. In addition, lane points on different lanes are also connected. An obvious case shown in Figure 5.6 is that the last lane point of a lane is connected with the first lane point of its successor lane. Also for two parallel lanes, the lane points are connected to each other if a legal lane switch could be made. Figure 5.6 demonstrates a possible cost configuration for the edges connecting lane points. We could set the cost of edges connecting lane points within the same lane to be 1. The cost of connecting to a right turn lane is set to be 5, and connecting to a left turn lane costs 8. Cost of moving along lane points is 2 within right turn lanes and 3 within left turn lanes. To emphasis the cost of switching lanes, cost of an edge connecting lane points in two different parallel lanes is set to be 10.

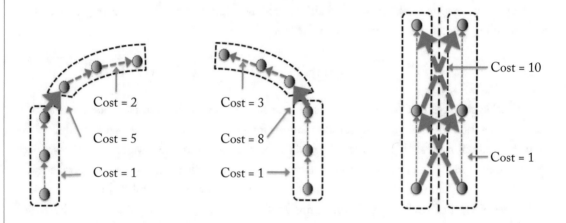

Figure 5.6: Costs of edges connecting lane points under scenarios of Right Turn, Left Turn, and Switch Lanes.

Under the cost configuration of Figure 5.6, we compare two different routes in the same road graph shown in Figure 5.7, both from the same source lane point (point A) to the same destination lane point (point B). Route 1 starts from Lane 1, and remains straightforward (Lane 4) at the bottom-left intersection. Then at the top-left intersection, it makes a right turn (Lane 5), and then keeps straight following Lane 10 and Lane 11, and finally reaches the destination via Lane 12; Route 2 also starts from Lane 1, but takes the right turn (Lane 2) at the bottom-left intersection and enters Lane 3. Then it makes a parallel lane switch to Lane 6 and takes a left turn (Lane 7 at the bottom-right intersection, which connects to Lane 8). At the top-right intersection, it follows the right turn into Lane 9 and also enters Lane 12 to reach the same destination B as route 1. Even though the total length of Route 2 might be smaller than Route 1's total length, the Routing module will prefer Route 1 under the cost configuration of Figure 5.6. Assuming that the cost of the edge connecting two lane points whose endpoint stays on a non-turn lane is 1, the total cost of Route 1 is 23, while total cost of Route 2 is 45.

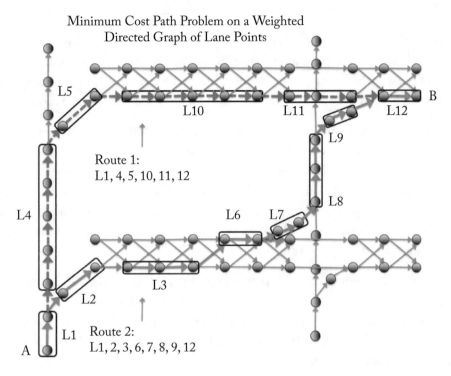

Figure 5.7: Routing as a minimum cost path problem on lane points connected weighted graph.

5.3.2   TYPICAL ROUTING ALGORITHMS

With the Routing problem in autonomous driving formalized in Section 3.1, we now introduce two typical routing algorithms: Dijkstra [15] and A*[16].

Dijkstra Algorithm on Autonomous Vehicle Routing

Dijkstra is a very common shortest path algorithm in graph theory. Proposed by Edsger W. Dijkstra in 1959 [15], the algorithm solves the shortest path from a source node to a destination node on a weighted graph. Applying the algorithm on the formalized lane point-based routing, the detailed algorithm is described as follows.

1. From the HD map interface, read the connected lane graph information data within a radius range, sample lane points on the lanes, and construct the lane point connected graph as described in Section 5.3.1. The lane point closest to location of the autonomous vehicle (as the "master vehicle") is set as the source node, and the lane

point closest to the destination is set as destination node. Set the source node costs to all the other node as infinity (*inf*), which indicates that cost between source and destination node is infinity. Cost to the source node itself is 0.

2. Set the current node to the *Source Lane Point*. Label all the other lane points as unvisited and put them into a set (*unvisited set*). In the meantime, we maintain a map (*prev_map*), which maps a lane point to its *Predecessor* lane point. This map stores the mapping of a visited lane point to its *Predecessor* lane point on the shortest path.

3. From the *Current Lane Point*, consider all the adjacent lane points which are unvisited and compute the *tentative distance* to reach these unvisited lane points. For example, the current lane point X is labeled distance of 3, and the distance between X and Y is 5. Then the *tentative distance* to Y could be 3 + 5 = 8. Compare this tentative distance to Y's current *labeled distance*. If Y's current labeled distance is smaller, we keep it. Otherwise replace Y's current labeled distance as this new tentative distance and update the *prev_map* accordingly.

4. For all the unvisited lane points connected to the *Current Lane Point*, repeat the process in Step 3. When all the adjacent lane points of the *Current Lane Point* have been processed, set the current node as visited and remove it from the *unvisited set*. The lane points which are removed from the unvisited set will no longer be updated for their labeled minimum distances.

5. As long as our destination point is still in the *unvisited set*, keep extracting lane point from the unvisited set and make it the current node and repeat Steps 3 and 4. The process will end when our *Destination Lane Point* has been removed from the *unvisited set* or the lane point node with minimum *tentative distance* in the *unvisited set* is infinity, which indicates that within a certain radius there is no possible way to reach the nodes remaining in the *unvisited set* from the *Source Lane Point*. The latter case means no routes available or a routing request failure, in which case the routing module needs to notify the downstream module or tries to re-route by loading the road graph information with a larger radius range.

6. If a shortest distance path has been found, construct and return the actual shortest path from the *prev_map*.

The pseudo code implementing the Dijkstra algorithm on the weighted directed graph of lane points is shown in Figure 5.8. Lines 2–16 is the typical Dijkstra algorithm that constructs the table of minimum tentative distances between lane points. Then from lines 17–22, based on the mapping of the minimum tentative distance corresponding *Predecessor* lane point, the algorithm constructs the actual shortest path by traversing the *prev_map* one by one. The output of the algorithm is a sequential list of lane points, which we cluster into actual routing lane segments as {(*lane*, *start_position*, *end_position*)} in line 23.

```
1 function Dijkstra_Routing(LanePointGraph(V,E), src, dst)
2     create vertex set Q
3     create map dist, prev
4     for each lane point v in V:
5         dist[v] = inf
6         prev[v] = nullptr
7         add v to Q
8     dist[src] = 0
9     while Q is not empty:
10            u = vertex in Q s.t. dist[u] is the minimum
11            remove u from Q
12            for each connected lane point v of u:
13                candidate = dist[u] + cost(u, v)
14                if candidate < dist[v]:
15                    dist[v] = candidate
16                    prev[v] = u;
17        ret = empty sequence
18        u = dst
19        while prev[u] != nullptr:
20                insert u at the beginning of ret
21                u = prev[u]
22        insert u at the beginning of ret
23        merge lane point in ret with same lane id and return the merged sequence
```

Figure 5.8: Dijkstra implementation of routing based on weighted directed graph of lane points.

The lane point graph we constructed has V nodes and E edges. Using a minimum priority queue to optimize extraction of the node with minimum distance at line 10, the running complexity of Dijkstra-based routing could be $O(|E| + V log|V|)$.

A* Algorithm on Autonomous Vehicle Routing

Another very popular routing algorithm which could be used for autonomous vehicle routing is the A* algorithm. A* is a heuristic-based search algorithm. Like the breadth first search (BFS) and depth first search (DFS), A* also searches the space according to some metrics. It could be viewed as a "merit-based" or "best-first" searching algorithm.

A* will maintain a set of nodes, called *openSet*, which contains potential nodes to expand for searching. In every loop, A* will extract the node with minimum cost to expand for searching. The cost of the node $f(v)$ to extract has two contributing parts: $f(v = g(v + h(v)$. First, in the search tree of A*, every node will have a cost representing the cost of getting from the source node to this node, denoted as $g(v)$; meanwhile, every node v has a heuristic cost denoted as $h(v)$. The heuristic cost

$h(v)$ represents an estimate of the minimum cost of getting from the current node to the destination node. When the heuristic $h(v)$ satisfies certain properties, A* is guaranteed to be able to find the minimum cost path from source to destination. In every loop of the A* searching process, the node with the minimum cost of $f(v = g(v + h(v)$ will be expanded until the destination node is expanded. The detailed algorithm implementation is as Figure 5.9, in which the function *reconstruction_route* in line 11 is similar as the routing reconstruction part in the Dijkstra algorithm.

```
 1 function AStar_Routing(LanePointGraph(V,E), src, dst)
 2     create vertex set closedSet        // set of already visited nodes
 3     create vertex set openSet          // set of nodes to be expanded
 4     insert src into openSet
 5     create map gScore, fScore with default value inf
 6     create prev_map with default value nullptr
 7     fScore[src] = h(src, dst)
 8     while openSet is not empty:
 9         current = the node v in openSet s.t. fScore[v] is minimum in openSet
10         if current = dst
11             return reconstruction_route(prev_map, current)
12         remove current from openSet
13         insert current into closedSet
14         for each neighbor u of current:
15             if u is in closedSet:
16                 continue; // ignore the neighbor who has already been evaluated
17             candidate_score = gScore[current] + h(current, u)
18             if u not in openSet:        // discovered a new node
19                 insert u into openSet
20             else if candidate_score >= gScore[u]:  // this is not a better path
21                 continue;
22             prev[u] = current
23             gScore[u] = candidate_score
24             fScore[u] = gScore[u] + h(u, dst)
```

Figure 5.9: A* Algorithm on autonomous vehicle routing.

As a heuristic-based search algorithm, when the $h()$ function satisfies the admissible [16] property, meaning that the minimum cost estimation $h(v,dst)$ never exceeds the actual minimum cost, then the A* algorithm could always find the minimum path. When this property is not satisfied, there is no guarantee that A* will find the minimum path. Under the scenario of autonomous vehicle routing, with the Lane Point connected graph, one way to define the heuristic measure between any two lane points A and B is: $h(u,v) = dist(u,v)$. The $dist()$ represents the Mercator [17] distance for two lane points under the earth geographical coordinate system. A* as a best-first search algorithm, could be viewed as an extension of the Dijkstra algorithm. Vice versa, the Dijkstra algorithm could be viewed as a special case of A* where the heuristic $h(u,v) = 0$.

5.3.3   ROUTING GRAPH COST: WEAK OR STRONG ROUTING

In practice, the choice of algorithm is usually not as important as the configuration of costs in autonomous vehicle routing. How to adjust the costs between the lane points, as demonstrated in Section 5.3.1 is the critical factor in building a working routing module. For example, if we know from dynamic traffic information that a road is very crowded, then we could put high costs on edges connecting the lane points belonging to this road such that the routing could avoid this congested lane; similarly, if there is traffic control for certain roads, we could also set the cost of connecting to lane points on the roads to be high enough (like infinity) such that these lanes could be less preferred or avoided in the search algorithm. In addition, the costs between lane points could be dynamically adjusted to reflect certain lane preferences over other lanes.

Considering the fact that actual road graph information data is very large, the routing module could usually pre-load the road graph and construct the lane point graph in an ad-hoc fashion. If route to destination is not available given a small radius of road graph loaded, the routing module could re-load a larger radius of road graph data, re-construct the Lane Point graph and re-compute the routing. There are typically two types of routing requests: the first type is usually when the autonomous vehicle starts its journey and the passenger sets the source and the destination by sending a routing request; the second type of routing is usually initiated by downstream modules (such as Behavior Decision or Motion Planning).

Here we introduce the notions of *strong routing* and *weak routing*. Strong routing implies that the downstream modules will strictly follow the results of the routing module. This means in the sense of lane by lane, the decision and planning modules will do their best to stay on the routing designated lanes. When they could not possibly follow the lanes of routing, they will send a re-routing request, as described in the second type of routing requests. For weak routing, downstream modules will not strictly follow the routing result under certain necessary scenarios. Weak routing will lead to the actual lane sequence taken to be different from the routing result, or in other words, different autonomous vehicle behavior. Consider a scenario where the routing output indicates the autonomous vehicle needs to stay on the current lane. Let's assume that there is a slow-moving vehicle on the current lane in front of the autonomous vehicle. With strong routing, the autonomous vehicle will reduce speed and just *follow* the slow vehicle in front. However, with weak routing, the autonomous vehicle might take the action of switching to the adjacent parallel lane, head-pass the slow vehicle, and then merge back to the previous lane as most human drivers will do. In dependent of strong or weak routing, whenever emergency accidents happen or there is necessity to perform urgent maneuvers, the downstream modules will act with the overarching principle of *safety first*, and a re-routing request will usually be sent in such cases.

5.4 CONCLUSIONS

We discussed the prediction and routing modules in this chapter, neither of which are within the traditional planning and control concept of modules. However, with our proposed broader sense of planning and control framework, both Prediction and Routing generate inputs for the traditional motion planning. Therefore, this is why we incorporated them within the broader planning and control framework. Traffic prediction is abstracted into a two-layered behavior classification and trajectory generation problem, and our proposed routing here is a lane-level routing which navigates our autonomous vehicle to its destination via the routed lane sequences. With the predicted trajectory and routing ready, we will describe typically traditional planning and control modules, including behavioral decision, motion planning, and feedback control in Chapter 6.

5.5 REFERENCES

[1]   Paden, B., Cap, M., Yong, S. Z., Yershow, D., and Frazzolo, E. 2016. A survey of motion planning and control techniques for self-driving urban vehicles. *IEEE Transactions on Intelligent Vehicles*, 1(1) pp. 33-55. DOI: 10.1109/TIV.2016.2578706. 89, 90, 91

[2]   Buehler, M., Iagnemma, K., and Sanjiv, S. (eds.). 2009. *The DARPA Urban Challenge: Autonomous Vehicles in City Traffic*. Springer Tracts in Advanced Robotics. DOI: 10.1007/978-3-642-03991-1. 90, 91

[3]   Montemerlo, M., Becker, J., Bhat, S., Dahlkamp, H., Dolgov, D., Ettinger, S., Haehnel, D., Hilden, T., Hoffmann, G., Huhnke, B., Johnston, D., Klumpp, S., Langer, D., Levandowski, A., Levinson, J., Marcil, J., Orenstein, D., Paefgen, J., Penny, I., Petrovskaya, A., Pflueger, M., Stanek, G., Stavens, D., Vogt, A., and Thrun, S. 2008. Junior: The Stanford entry in the urban challenge. *Journal of Field Robotics: Special Issue on the 2007 DARPA Urban Challenge*, 25(9), pp. 569-597. 90

[4]   Gindele, T., Brechtel, S., and Dillmann, R. 2010. A probabilistic model for estimating driver behaviors and vehicle trajectories in traffic environments. In *Proceedings of the 13th International IEEE Conference on Intelligent Transportation Systems (ITSC)*, Madeira Island, Portugal, pp. 1625–1631. DOI: 10.1109/ITSC.2010.5625262. 89

[5]   Aoude, G. S., Desaraju, V.R., Stephens, L. H., and How, J.P. 2012. Driver behavior classification at intersections and validation on large naturalistic data set. *IEEE Transactions on Intelligent Transportation Systems,* 13(2), 724–736. DOI: 10.1109/tits.2011.2179537. 89

[6]   Lefevre, S., Gao, Y., Vasquez, D., Tseng, H.E., Bajcsy, R., and Borrelli, F. 2014. Lane keeping assistance with learning-based driver model and model predictive control. In

Proceedings of the 12th International Symposium on Advanced Vehicle Control, Tokyo, Japan. 89

[7]   Gadepally, V., Krishnamurthy, A., and Ozguner, U. 2014. A framework for estimating driver decisions near intersections. *IEEE Transactions on Intelligent Transportation Systems*, 15(2), pp. 637–646. DOI: 10.1109/TITS.2013.2285159. 89

[8]   Gadepally, V., Krishnamurthy, A., and Ozgüner, U. 2016. *A Framework for Estimating Long Term Driver Behavior*. arXiv, 2016; arXiv:1607.03189. 19. 92

[9]   Bonnin, S., Weisswange, T.H., Kummert, F., and Schmuedderich, J. 2014. General behavior prediction by a combination of scenario-specific models. *IEEE Transactions on Intelligent Transportation Systems*, 15(4), pp. 1478–1488. DOI: 10.1109/TITS.2014.2299340. 92

[10]   Kumar, P., Perrollaz, M., Lefevre, S., and Laugier, C. 2013. Learning-based approach for online lane change intention prediction. In *Proceedings of the IEEE Intelligent Vehicles Symposium* (IV 2013), Gold Coast City, Australia, pp. 797–802. DOI: 10.1109/IVS.2013.6629564. 92

[11]   Hsu, C.W., Chang, C.C., and Lin, C.J. 2003. *A Practical Guide to Support Vector Classification*. Department of Computer Science, National Taiwan University. 96

[12]   Krizhevsky, A., Sutskever, I., and Hinton, G.E. 2012. Imagenet classification with deep convolutional neural networks. In *Advances in Neural Information Processing Systems* (pp. 1097-1105). 96

[13]   Medsker, L.R. and Jain, L.C. 2001. Recurrent neural networks. *Design and Applications*, 5. 97

[14]   Sak, H., Senior, A., and Beaufays, F. 2014. Long short-term memory recurrent neural network architectures for large scale acoustic modeling. In *Fifteenth Annual Conference of the International Speech Communication Association*. 97

[15]   *Dijkstra's Algorithm*. https://en.wikipedia.org/wiki/Dijkstra's_algorithm. 103

[16]   *A* Algorithm*. http://web.mit.edu/eranki/www/tutorials/search/. 103

[17]   *Earth Coordination System. https://en.wikipedia.org/wiki/Geographic_coordinate_ system*. 106

[18]   Rasmussen, C.E. 2006. *Gaussian Processes for Machine Learning*. The MIT Press. 98

CHAPTER 6

Decision, Planning, and Control

Abstract

*In this chapter, we continue with our general discussion of the planning and control modules by expanding on the concepts of behavior **decision**, motion **planning**, and feedback **control**. Decision, planning, and control are the modules that compute how the autonomous vehicle should maneuver. These modules constitute the traditional narrow concept of planning and control. While they all solve the same problem of how the autonomous vehicle should behave, they operate at different levels of the problem abstraction.*

6.1 BEHAVIORAL DECISIONS

The *behavior decision* module acts as the "co-driver" in the general autonomous vehicle planning and control modules. It is the module where most of the data sources are consumed and processed. Data sources fed to the decision module includes, but are not limited to, information about the autonomous vehicle itself including location, speed, velocity, acceleration, heading, current lane info, and any surrounding objects information within a certain radius. The mission of behavioral decision module is to compute the behavioral level decision given all these various input data sources. These input data sources may include the following.

1. **The routing output:** A sequence of lanes along with the desired starting and ending position (where to enter and leave along the lane).

2. **The attributes about the autonomous vehicle itself:** Current GPS position, current lane, current relative position given the lane, speed, heading, as well as what the current target lane is, given the autonomous vehicle location.

3. **The historical information about the autonomous vehicle:** In the previous frame or cycle of behavioral decision, what is the decision output, e.g., was it to follow, stop, turn, or switch lanes?

4. **Obstacle information around the autonomous vehicle:** This would pertain to all the objects within a certain range of the autonomous vehicle. Each perceived object

contains attributes such as present lane, speed, heading, as well as their potential intentions and predicted trajectories. Object and attribute information are mostly produced by the *perception* and *prediction* modules.

5. **Traffic and map objects information:** The lanes and their relationships as defined by the HD map. For example, Lane 1 and Lane 2 are adjacent, and it is legal to make a lane switch. Then what is the legal range for switching lanes? Another example is, when we finished a straight lane and need to enter into a left-turn lane, is there a traffic light or stop-sign or pedestrian cross-walk at the connection of these two lanes? This kind of information comes from the mapping module as well as from the perceived dynamic traffic signs (e.g., traffic light green or red).

6. **Local traffic rules:** For example, the city speed limit or if it is legal to make a right-turn at the red light.

The goal of the *decision* module is to leverage all these pieces of information and make effective and safe decisions. It is easy to see that the decision module is where all the data sources are considered. Due to the heterogeneous characteristic of these data sources, and the various different local traffic laws, it is quite difficult to formulate the behavioral decision problem and solve it with a uniformed mathematical model. It is more suitable to use advanced software engineering concepts and design a traffic rule-based system to solve this problem. In fact, an advanced rule-based behavior decision system can be found in many successful autonomous driving systems. For instance, in the DARPA challenge, Stanford's autonomous driving system "Junior" [1] utilizes a Finite-State-Machine (FSM) with cost functions to deterministically compute the trajectory and behavior of the vehicle. Similarly, the CMU autonomous driving system "Boss" [2] computes the space gaps between lanes and utilizes such information together with pre-encoded rules to trigger the lane switching behavior. Other systems such as Odin and Virginia Tech [3] also used rule-based engines to decide on the behavior of their vehicles. With increased research efforts in autonomous driving decision and planning systems, Bayesian models are becoming more popular in modeling the autonomous vehicle behavior and have been applied in recent research works [4, 5]. Among the Bayesian models, Markov Decision Process (MDP) and Partially Observable Markov Decision Process (POMDP) are the most widely applied methods in modeling autonomous driving behavior.

Even though academic researchers often prefer non-deterministic Bayesian model approaches, we do believe that, in practical industrial systems, rule-based deterministic decision systems still have a key role to play. We will describe a rule-based approach with a typical scenarios in this section.

The rule-based approach we introduce is based on the principle of *Divide and Conquer* to decompose the surrounding environment into layered scenes and approach each layer individually.

In fact, we believe that in an actual autonomous driving production system, a rule-based system will even be safer and more reliable given its simplicity. Imagine how human drivers drive from point A to point B via a fixed route. The traffic rules are always fixed. More importantly, given the surround- ing environment including nearby vehicles, pedestrians and traffic signs, if we apply traffic rules together with our intention of where to go, the behavioral output, or how the human driver should re-act, is usually constrained within a very limited number of behavioral choices, or even clearly specified by traffic rules. For example, in California, if a vehicle wants to cross a four-way stop sign intersection, it should first stop for 3 seconds, yield to any other vehicle that has the right of way before proceeding. The whole series of actions is determined with clear consideration of the surrounding objects and it could be naturally modeled in a deterministic fashion. Even though there might be unexpected conditions which may lead to a violation of certain traffic rules, the principle of *safety first* to avoid collision could also be deterministically enforced.

6.1.1   MARKOV DECISION PROCESS APPROACH

A Markov Decision Process (MDP) is defined by the following five element tuple (S, A, P_a, R_a, γ), where:

1. S represents the state space of the autonomous vehicle. The state space should consider the location of the autonomous vehicle along with map elements such as lanes. One can divide the surrounding square of the vehicle into grids of fixed length and width. Considering different road map objects, such as lanes and reference lines, we could also create spaces concerning different combination of map objects, such as the current and adjacent lanes where the autonomous vehicle is located;

2. A represents the behavioral decision output space, which is a fixed set of all possible behavioral actions which can be taken: for instance, a decision state could be to *Follow* a vehicle on the current lane, *Switch Lane* to an adjacent parallel lane, *Turn Left/ Right*, *Yield*, or *Over-take* a crossing vehicle at a junction, *Stop* for traffic lights and pedestrians, etc.;

3. $P_a (s, s') = P(s' | s, a)$ is the conditional probability to reach state s', given that the vehicle is currently at state s and takes action a;

4. $R_a (s, s')$ is the reward function fo transforming from state s to state s' by taking action a. The reward is a synthetized measure of how we evaluate such state transformations. Factors that should be considered and represented in the reward include: safety, comfort, ability to reach the destination, and difficulty for the downstream motion planning to execute; and

5. γ is the decay factor for thereward. The reward at the current time has a factor of 1 while the reward for the next time frame will be discounted by a factor of γ. Accordingly, the reward for t time frames in the future will be deemed as γ^t for the time being. The decaying factor guarantees that the same amount of reward will always be more valuable for the time being than in the future. Note that the meaning of the decay factor is that shorter term rewards is more valuable than longer term rewards. For example, it is better to reach a certain desire speed (like user-defined cruising speed) as soon as possible;

With the formal MDP setting, the problem that behavioral decision needs to solve, is to find an optimal *policy* π to perform the mapping $S{\rightarrow}A$. In other words, given a state s, the policy must compute a behavioral decision a = $\pi(s)$. When the policy has been determined, the whole MDP could be viewed as a Markov Chain. The behavioral decision policy π is to optimize the accumulated rewards from the present time to the future. Note that if the reward is not deterministic but a random variable, then the policy will optimize the expected accumulated rewards. Mathematically, the accumulated reward to maximize is written as:

$$\sum_{t=0}^{\infty} \gamma^t R_{at}(s_t, s_{t+1}),$$

where action a is the policy output $a = \pi(s)$. The method to find this policy is usually based on *Dynamic Programming*: ssume that the state transition probability matrix P and reward distribution matrix R are known, the optimal policy solution could be obtained by iterating on the computing and storing the following two state arrays:

$$\pi(s_t) \leftarrow \underset{a}{\text{argmax}} \left\{ \sum_{s_{t+1}} P_a(s_t, s_{t+1})(R_a(s_t, s_{t+1}) + \gamma V(s_{t+1})) \right\}$$

$$V(s_t) \leftarrow \sum_{s_{t+1}} P_{\pi(s_t)}(s_t, s_{t+1})(R_{\pi(s_t)}(s_t, s_{t+1}) + \gamma V(s_{t+1})).$$

$V(s_t)$ represents the accumulated future rewards discounted at the current time, and $\pi(s_t)$ represents the policy that we want to search for. The solution is based on repeatedly iterating between possible state pairs (s, s'), until the above two state arrays converge [6, 7]. Furthermore, in Bellman's value iteration algorithm, there is no need to explicitly compute $\pi(s_t)$. Instead, $\pi(s_t)$ related computation could be incorporated into computation of $V(s_t)$, which leads to the following single step "Value Interation":

$$V_{i+1}(s) \leftarrow \underset{a}{\text{max}} \left\{ \sum_{s'} P_a(s, s')(R_a(s, s') + \gamma V_i(s')) \right\},$$

where i is the iteration step. At step $i = 0$, we make an initial guess of $V_0(s)$. $V(s)$ is updated at each step until convergence. There are various methods for applying MDP to autonomous vehicle control but it would be outside the scope of this book to examine different MDP autonomous driving decision implementation details. Instead, the interested reader is referred to [6, 7] for a better understanding of how to design state spaces, action spaces, state transitions, and the reward function implementation.

We want to emphasize here a few factors which must be considered when designing the reward function Ra (s, s'), since it is one of the most critical elements in building a working MDP decision system. A good reward function in an MDP-based decision module should exhibit the following characteristics.

1. **It should be able to reach each the destination:** *Incentivize* the autonomous vehicle to follow the routing module output route to reach the destination. In other words, if the action chosen by the policy, i.e., $a = \pi(s)$, leads the vehicle to diverge from the route, some "punishment" should be meted out. Conversely, a reward should be given out for actions that follow the route.

2. **It should be safe and collision free:** If the state is based on $N \times N$ equal square grids centered on the vehicle, then any decision to move to another grid where collision might occur should be negatively rewarded. Movements to grids with a lower collision possibility or larger distance to collision likely grids should be rewarded.

3. **The ride should be comfortable and smooth:** A comfortable journey mostly indicates few or no sharp maneuvers. Further, a limited amount of abrupt maneuvers helps the downstream modules to smoothly execute most of the decisions. For example, a smooth change of speed should have a higher reward than a sharp acceleration or deceleration.

At this point, the reader can probably sense how delicate the job is to build a working MDP-based decision system when considerations of state space design, action space design, transition probability matrix, reward function, etc. must be balanced against one another.

6.1.2   SCENARIO-BASED DIVIDE AND CONQUER APPROACH

The key idea is to apply the principle of *Divide and Conquer* to decompose the vehicle's surroundings into scenarios. In each scenario, the corresponding rule will be applied individually to the objects or elements in the scenarios to compute an *Individual Decision* for each object, and then a *Synthetic Decision* for the vehicle itself is computed by consolidating all the individual decisions. For example, when we are trying to stay in our current lane, we will first figure out whether there is a lead vehicle we care about. The corresponding rule in this example is how to determine which vehicle qualifies as the lead vehicle when we want to stay in our ego lane. The rule to determine whether there is a lead vehicle in our ego lane is based on factors like our intension (stay-in-lane), our preferred speed, and the surrounding obstacles. When we have determined a lead vehicle, the individual decision tag for this lead vehicle is a FOLLOW. And the synthetic decision for our ego vehicle is also FOLLOW. Essentially, based on our intention and the environment, we provide tags to obstacles as individual decisions. And the important individual decisions form our description of Synthetic Decisions from the ego point-of-view.

Synthetic Decision

Synthetic Decision	Parametric Data
Cruise	➤ Current lane ➤ Speed limit of the current lane
Follow	➤ Current lane ➤ *id* for the vehicle to follow ➤ Speed to reach minimum of current lane speed limit and speed of the vehicle to follow ➤ Not exceeding 3 m behind the vehicle in front
Turn	➤ Current lane ➤ Target lane ➤ Left or right turn ➤ Speed limit for turning
Change Lane	➤ Current lane ➤ Target lane ➤ Change lane by overtaking and speed up to 10 m/sec ➤ Change lane by yielding and speed down to 2 m/sec
Stop	➤ Current lane ➤ *id* for any object to stop, if any ➤ Stop by 1 m behind the object to stop

Figure 6.1: Synthetic decision with its parameters in behavioral decision.

The notion of *Synthetic Decision* is regarding how the autonomous vehicle itself should behave. It is the top-level behavioral decision. Example synthetic decisions include: remain in the current lane to follow another vehicle, switch lane to an adjacent parallel lane, or stop by a certain stop-line as required by a traffic sign or light. As the top-level decision behavior, its possible output space, along with its definitions, must be consistent and shared with the downstream motion planning module. In addition, to help motion-planning to come out with the planned trajectory, the synthetic decision is always companied with parameters. Figure 6.1 lists a few synthetic decision definitions as well as their possible parameters. Consider when the synthetic decision at the current time frame is to *Follow*. The output command to the motion planning module is not only the behavioral *follow* command, but also parameters such as: the *id* of the vehicle to follow on the current lane, the suggested speed to follow (which is usually that of the vehicle in front, or the lane speed limit, whichever is lower), and suggested distance to keep while following (for example, 3 m behind the vehicle in front). This allows the downstream motion planning to utilize these parameters as constraints, to compute a smooth and collision-free trajectory.

Individual Decision

In opposition to *synthetic decisions* are *individual decisions*. As we mentioned, a synthetic decision is a comprehensive consolidated decision for the vehicle itself after considering all the information including all the road objects. We thus propose to explicitly produce an *individual decision* for each element in our surrounding world. The object accompanied by an individual decision could be an actual perceived obstacle on the road or just a logical map object such as the stop-line associated with a traffic light or a pedestrian crosswalk. Actually, in our design shown in Figure 6.4, the logic of scenario division takes place first. Individual decisions for objects are then computed and associated with each object in all the scenarios. Only after individual decisions have been made about all the objects can the final synthetic decision come. It is a consolidation of all the individual decisions. Like synthetic decision, the individual decisions also come with parameters. These individual decisions are not only necessary pre-requisites to compute the synthetic final decision, but also transmitted to downstream motion planning modules to facilitate trajectory planning. For example, when we want to change to the right-neighbor lane, we consider the vehicles in our right-neighbor lane, as well as vehicles in our ego lane. For the vehicles in our right-neighbor lane, if the spatial and temporal distances are large enough, we may provide OVERTAKE or YIELD individual decision tags to them. And before we actually change lanes, we still need to keep a distance to our ego lane lead vehicle if there is one. So, there might also be an individual decision for the ego lane lead vehicle. That individual decision for ego lane lead vehicle might be FOLLOW or even IGNORE. With all these necessary obstacles considered by tagging them with individual decision, we might stick to our intention of change lane, where we will put CHANGE LANE as our synthetic decision. This CHANGE LANE synthetic decision was computed by consolidating the individual decisions.

At this point, the reader might wonder why these individual decisions are also sent to downstream modules. Would not it be sufficient to convey the final synthetic decision given that we only plan the action for the vehicle itself? As it turns out, our experience has shown that sending both the synthetic final decision and its supporting individual decisions is beneficial for the downstream motion-planning task. Since these individual decisions serve as projections of the synthetic decision in a consistent way, motion planning will have much more reasonable and explicit constraints and hence the optimization problems of motion-planning can be better formalized when furnished with individual decisions. In addition, debugging efficiency will be greatly improved with individual decisions.

Figure 6.2 lists a few typical individual decisions and their parameters. For example, if the individual decision for an object X is to *overtake*, then the parameters associated with this overtake decision will possibly include the time and distance needed to overtake the object X. The distance parameters should include the minimum empty space which should be in front of the object X, while the time parameter is the minimum time corresponding to how long this empty space should be present, given the respective speeds of the autonomous vehicle and of the object X. Note that such "overtake or yield" individual decisions only exist when an object's predicted trajectory intersects the planned trajectory of the vehicle. Typical examples of yielding/overtaking an object include scenarios at junctions. We will use an example at a junction to illustrate how exactly we divide the surroundings into layered scenarios, apply specific rules to obtain individual decisions, and finally consolidate them as synthetic decision output.

Individual Decision		Parametric Data
Vehicle	Follow	➢ *id* for the vehicle to follow
		➢ Speed to reach for following the vehicle
		➢ Distance to keep for following the vehicle
	Stop	➢ *id* for the vehicle to stop
		➢ Distance to stop behind the vehicle
	Attention	➢ *id* for the vehicle to stop
		➢ Minimum distance to keep while paying attention to the vehicle
	Overtake	➢ *id* for the vehicle to overtake
		➢ Minimum distance to keep for overtaking
		➢ Minimum time gap to keep for overtaking
	Yield	➢ *id* for the vehicle to yield
		➢ Minimum distance to keep for yielding
		➢ Minimum time gap to keep for yielding
Pedestrian	Stop	➢ *id* for the pedestrian to stop
		➢ Minimum distance to stop by the pedestrian
	Swerve	➢ *id* for the pedestrian to swerve
		➢ Minimum distance to keep while swerving around

Figure 6.2: Individual decisions with parameters in behavior decision module.

Scenario Construction and System Design

The computation of individual decisions is dependent on the construction of *scenarios*. Here one could simply think of the scenarios as a series of relatively independent partitions for the surrounding world of the vehicle. We opted to partition the surrounding world in a layered structured fashion. This means that scenarios belong to different layers and the scenarios within layer are independent. A deeper layer of scenarios could leverage any computation result or information from the shallower layers. Objects usually belong to only one scenario. The idea behind this structured layered scenario partitioning of the world remain the principle of *divide and conquer*: we first focus on independent small *worlds*, i.e., scenarios, and solve the problem of computing the decisions within that small world. While computing the individual decisions in each independent scenario in the same layer, the routing intention (where the autonomous vehicle should go) and the previous layered computation results are shared. After obtaining the individual decisions, a synthetic decision is consolidated with a set of rules. Figures 6.3(a) and 6.3(b) show two examples of how we partition into scenarios and compute behavioral decisions.

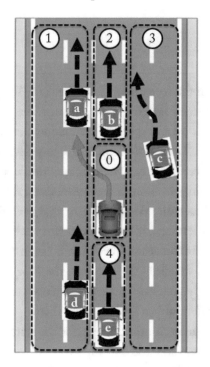

Synthetic Decision:
  Switch lane from the current lane to the left lane: yield to vehicle **a**, overtake vehicle **d**, and pay attention to vehicle **b** in the current lane.

Scenarios and Individual Decisions:
  1. Master Vehicle
  2. Left Lane(s); Overtake **d** and yield to **a**
  3. Front Vehicle(s): Pay attention to **b**
  4. Right Lane(s): Ignore **c**
  5. Rear Vehicle(s): Ignore **e**

Figure 6.3: (a) Layered scenarios while performing a lane switch.

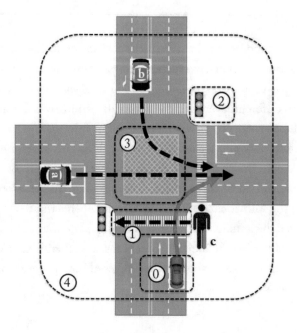

Synthetic Decision:
  Stop by the crosswalk stop-line and
  wait for pedestrian **c** to cross

Scenarios and Individual Decisions:
  First Layer Scenarios
    1. Master Vehicle
    2. Crosswalk: Stop for pedestrian **c**
    3. Traffic Light: Red light turn right,
        yield any through/turn traffic
    4. Keep Clear Zone Ignore
  Second Layer Scenarios
    5. Junction Scenarios: Based on
        Scenarios 1, 2, 3

Figure 6.3: (b) Layered scenarios while at junction.

In Figure 6.3(a), there are two vehicles **a** and **d** in the scenario of "Left Lane(s)". The intention of the autonomous vehicle is to switch from the current lane to its left adjacent lane as specified by the routing output. Considering the relative position and speed of the autonomous vehicle regarding vehicle **a** and **d**, the computational result of the *Left Lane(s)* scenarios is to yield vehicle **a** and overtake vehicle **d**, which means to switch lane between these two vehicles; meanwhile, another scenario of *Front Vehicle(s)* depicts the small world of things in front of the autonomous vehicle itself, and this scenario is independent of the *Left Lane(s)* scenario. We should note that, even though the intention of autonomous vehicle is to switch to the left lane, it is still important not to ignore anything ahead in the current lane. Therefore, the individual decision for vehicle b in the *Front Vehicle(s)* scenarios is to keep an appropriate distance from vehicle b. We also have the *Rear Vehicle(s)* and *Right Lane Vehicle(s)* scenarios. However, given that the predicted trajectories of objects in these scenarios do conflict with our planned trajectory, we will be able to safely ignore them

Scenarios in Figure 6.3(a) do not depend much on each other, except for the status of the *Master Vehicle* information which is shared among all these scenarios. In Figure 6.3(b), we show a complex case with many layers of scenarios. The *Master Vehicle* scenario is a special one whose information will be shared to be utilized by other scenarios. The first layer of scenarios includes *Front/Rear Vehicle(s)*, *Left/Right Lane Vehicle(s)*, and traffic sign area-related scenarios such as *Traffic Light* and *Crosswalk*. More complex *composite* scenarios could be built on top of the first layer scenarios by using them as elements. As shown in Figure 6.3(b), the four-way intersection scenario is based on the scenarios of *Crosswalk*, *Traffic Light*, and *Master Vehicle*. Besides these membership scenarios, vehicles a and b belong to the four-way intersection scenario itself as they reside in lanes under the concept coverage of *junction*. Regarding membership scenario, we mean that the scenarios Crosswalk, Traffic Light, and Master Vehicle, are "members" of this complex scenarios. This called membership scenarios. Simply said, the junction scenario also applies here in this context. When implementing, you can think of it that every scenario has its own match to the real environment and they are somewhat independent.

Suppose the routing intention is to turn right and we currently see a red light as well as a pedestrian crossing the road. Traffic rules allow a right turn against the light but only after stopping and yielding to pedestrians or cross traffic. The individual decision regarding the crossing pedestrian will be to stop while the individual decisions for both vehicles **a** and **b** will be to yield. Consolidating these individual decisions, the synthetized decision for our autonomous vehicle itself will be to stop in front of the crosswalk-defined stop-line.

As described above, each individual scenario focuses on its own logic to compute the individual decisions for elemental objects within itself. Then, the behavioral decision module considers all the individual decisions for every object and comes up with a final synthetic decision for the autonomous vehicle itself by consolidating the individual decisions. Some special consideration must be given to the possibility that there are different or even conflicting individual decisions for the same object. For example, a vehicle could get two different individual decisions in two separate scenarios, one being *yield* and the other one being *overtake*. In general, the way we partition the surrounding world into scenarios is to naturally assign objects, either actual perceived objects or conceptual logical objects, into distinctive scenarios to which they belong. In most cases, an object will not likely appear in more than one scenario. However, we cannot completely rule out such possibilities. In fact, some of the scenarios do cover a small overlapped map area with robustness consideration. In detail, to make the system more robust, covered areas (or area of interests) for each scenario, might have some overlapping with each other. When such low probability cases do happen, a layer handles merging individual decisions for a safety and coherence check in the behavioral decision system (Figure 6.4). For example, imagine a vehicle in the same lane behind us in the process of changing from the current lane into the left lane, leading to its (temporary) "existence" in both the *Rear Vehicle(s)* and *Left Lane(s)*. Let us further assume that our autonomous vehicle also intends to switch to the left lane. The *Rear Vehicle(s)* scenario yields an attention individual decision while the *Left Vehicle(s)* scenario decides that we should *yield* to the vehicle. The individual decision merging layer will review these different individual decisions for each object and re-compute a merged final individual decision considering both safety and our autonomous vehicle intention. In this case, since our autonomous vehicle is also trying to switch to the left lane, we will then obtain a yield to the vehicle if we have already started the switching lane motion, or keep *paying attention* if we have not started switching lanes yet.

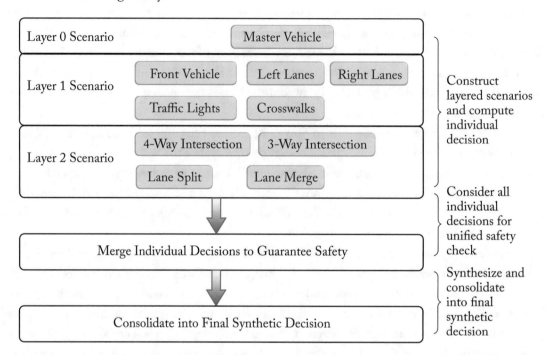

Figure 6.4: Architecture of a rule-based behavior decision system containing layered scenarios.

Overall, the system framework and logic process is shown in Figure 6.4. At the top are the layers focusing on the layered scenario construction, where information regarding our autonomous vehicle's intentions, mapping, and localization, and perceived surrounding world, are all utilized to building layers of independent scenarios. For each independent scenario's own business logic and the shared autonomous vehicle routing intention will determine the individual decisions for all the objects in the scenarios. After all the layered scenarios have finished their individual decision computation, the merging layer will double-check all the individual decisions and solve possible conflicts or inconsistency for any object. Finally, at the bottom, the final synthetic decision for the autonomous vehicle itself is computed by consolidating the merged and coherent individual decisions. This synthetic decision, along with the merged individual decisions, will be sent to the motion planning module, where a spatial-temporal trajectory for the autonomous vehicle will be planned for physical execution.

6.2   MOTION PLANNING

The *motion planning* module is directly downstream from behavioral decision. It is tasked with generating a trajectory and sending it to the feedback control module for physical vehicle control execution. The planned trajectory is specified and represented as a sequence of planned *trajectory points*. Each of these points contains attributes like location, time, speed, curvature, etc. The problem of autonomous vehicle motion planning can be viewed as a special case of general motion planning in robotics. In some sense, the motion planning problem for autonomous vehicle on a road is even easier than general motion planning in robotics since cars mostly follow the pre-existing road graph and move on a 2D plane. With the only control signals being throttle, brake, and steering wheel, the class of possible trajectories naturally exhibits certain characteristics such as smoothness and curvature constraints. Optimization is therefore comparatively easier than, say, planning a trajectory in 3D, with many more constraints (e.g., motion planning for a flying drone).

Since the DARPA urban challenge, motion planning in autonomous vehicles has been gradually developing as a relatively independent module. [4, 8] attempted to solve the problem of motion planning under certain conditions of urban driving and parking and there are also works on solving special motion planning problems such as found in [7]. [5, 9] list recent works on various aspects of motion planning.

Most motion planning work consists in optimizing a spatial-temporal trajectory within certain spatial-temporal constraints. A *spatial-temporal* trajectory consists of *trajectory points*. The attributes of each point include position, time, speed, acceleration, curvature, and even higher-order derivative of attributes such as curvature. These attributes are essential since the costs associated with these points constitute the optimization goal. The actual vehicle trajectory exhibits properties akin to spline trajectories. Therefore, motion planning could be formalized as an optimization problem for trajectories with certain common properties/constraints on the 2D plane. The vehicle trajectory is akin to spline trajectories since vehicles usually move smoothly and their trajectories are usually constrained in terms of curvature (kappa) and curvature change rate (kappa derivative). Spline trajectories usually have good properties in curvature and curvature change rate.

The two key elements in this optimization problem are *Optimization Object* and *Constraints*. Here the *Optimization Object* is usually represented as *costs* associated with different candidate solutions, and the goal of the optimization is to search for the one with minimum cost. The function that computes the cost is based on the following two key factors. First, as the direct downstream module of behavioral decision, the cost function must obey the upstream behavioral decision output. For example, the individual decision for a vehicle in front might be to follow and stay within a distance range relative to the rear of the front vehicle, then the planned trajectory must reach, but not exceed, the designated area specified by the individual decision. Also, the planned trajectory should be collision free, which means a minimum distance to any physical object has to be kept while computing the trajectory. Second, since we focus on autonomous driving on urban roads, the planned trajectory should be consistent with the shape of the road. This requirement essentially means that our vehicle should follow natural roads. All these aspects are represented in the design of cost functions. While the cost function design emphasizes obeying the up-stream behavioral decision and following routing directions, the constraints in the motion planning optimization problem is more about the constraints which the downstream feedback control can comfortably execute. For example, the curvature and second-order derivative of the curvature have constraints related to the steering wheel control. Similarly, with throttle as the control method for accelerating, the rate of acceleration changes is also limited.

As we mentioned at the beginning of the planning and control chapter, describing in detail all existing motion planning solutions would be beyond the scope of this book. Instead, we focus on two typical approaches, both extremely successful and proven. The first is based on a simpler version of the ideas presented in [9]. The problem of planning a spatial-temporal trajectory is partitioned into two problems to be tackled sequentially: *path planning* and *speed planning*. Path planning only solves the problem of computing trajectory shape on the 2D plane, given the behavior decision output and the cost function definition. The paths generated do not have any speed information and are merely splines of various shapes and lengths. Speed planning is based on the results of path planning and solves the problem of how the vehicle should follow a given trajectory. Compared with the proposed approach of [9] which simultaneously solves the problem of optimizing the spatial-temporal trajectory, our solution represents a clearer partition and formalization of the problem. Even though the proposal here might not necessarily find the optimal solution, industrial practices have shown that such a divide and conquer approach for motion planning is effective. Instead of partitioning the motion planning problem into path planning and speed planning, the problem is tackled by considering motion planning in two orthogonal directions, *longitudinal* (*s*-direction) and *lateral* (*l*-direction), as per the *SL-coordinate* system (which we will describe later in this section). The advantage of this approach over the first one is that the shape of the planned trajectory naturally takes speed into account, while the first approach may lead to selecting a trajectory shape inappropriate for the desired speed profile. This would be because trajectory and speed are separately optimized. While separately optimizing path and speed works well in urban low-speed autonomous driving, the second approach will be more suitable for higher speed scenarios such as highways.

6.2.1 VEHICLE MODEL, ROAD MODEL, AND SL-COORDINATION SYSTEM

Let us now discuss the mathematical concept of vehicle pose (pose can simply be treated as a vector of position and speed) and road-based *SL-coordination* system. A vehicle's pose is determined by: $\bar{x} = (x, y, \theta, \kappa, v)$ where (x, y) represents the position on the 2D surface, θ represents the direction of motion, κ is the curvature (the rate at which θ changes), and v represents the speed tangential to the trajectory. These pose variables satisfy the following relationship:

$$\dot{x} = v\cos\theta$$
$$\dot{y} = v\sin\theta$$
$$\dot{\theta} = v\kappa ,$$

where κ is a constraint given to the system. Considering a continuous trajectory generated by the vehicle, the direction along the path is the s-direction and the pose variables' relationship with the s-direction satisfy the following equations:

$$dx / ds = \cos(\theta(s))$$
$$dy / ds = \sin(\theta(s))$$
$$d\theta / ds = \kappa(s).$$

Note that here we have not placed any constraints on the relationship between κ and θ, meaning that the vehicle could change its curvature κ at any direction θ. However, in the actual control model, the relationship between curvature κ and direction θ is restrained, but this constraint is trivial in terms of the speed limit for urban road driving. It therefore does not have a significant impact on the practice and feasibility of the proposed motion planning algorithm.

The path-planning part of our proposed motion-planning algorithm heavily depends on the high-definition-map (HD-map), and especially on the *Center Line* of lanes in the map, which we refer to as *reference line*. Here a road lane is defined by its sampling function $r(s) = [r_x(s), r_y(s), r_\theta(s), r_\kappa(s)]$, where s represents the distance along the tangential direction of the path and *longitudinal* distance is the s-distance. Correspondingly, the *lateral* distance (l distance) represents the distance perpendicular to the s-direction. Consider a pose p under the (s, l) coordinate system and the corresponding pose under the (x, y) world coordinate system, the pose in the world coordinate system $p(s, l) = [x_r(s, l), y_r(s, l), \theta_r(s, l), \kappa_r(s, l)]$ satisfies the following relationships with the (s, l) coordinate pose:

$$x_r(s, l) = r_x(s) + l\cos(r_\theta(s) + \pi/2)$$
$$y_r(s, l) = r_y(s) + l\sin(r_\theta(s) + \pi/2)$$
$$\theta_r(s, l) = r_\theta(s)$$
$$\kappa_r(s, l) = r_\kappa(s)^{-1} - l)^{-1},$$

where κ_r is larger in the inner side of a turn than at the outside (meaning that κ_r increases with l for a constant s). As shown in Figure 6.5, near the x-axis at the origin, the lateral l increases with y. A Lane(k) with a constant width can be represented as a set of points along the longitudinal direction following the central reference line: $\{p(s, l) : \in R^+\}$. Such a lane coordinate system is formally refered to as the *SL-coordinate* system.

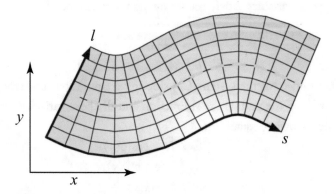

Figure 6.5: Lane-based SL-coordinate system under *XY* plane [9], used with permission.

6.2.2   MOTION PLANNING WITH PATH PLANNING AND SPEED PLANNING

Given the previously described road-based SL-coordinate system, we can solve the motion planning problem by first doing path planning and then speed planning. We define the vehicle *path* as a continuous mapping $\rho : [0,1] \to C$ from range $[0,1]$ to set of vehicle poses C = $\{\overline{x}'\}$. The initial pose of a planning cycle or frame is known as $\rho_1(0) = \rho_2(0) = q_{init}$ for path ρ_1 and ρ_1, which end separately at $\rho_1(1) = q_{end1}$ and $\rho_2(1) = q_{end2}$, as shown in Figure 6.6. The goal of path planning is to find a path, which starts from the initial pose, reaches a desired end pose, and satisfies certain constraints with minimum cost.

We search for the optimal cost path in a way similar to the way we used for computing routing, which is to place *sampling points* toward potential areas which the path might traverse. In detail, the candidate trajectories are made of piece-wise polynomial splines. To determine such splines, a start point and an end point are needed. In short, connecting the sampled points will generate a candidate trajectory. In Figure 6.6, we uniformly divide the lane into *segments* of equal *s* and *l* distances. The central point in each divided small (s_i, l_j) grid is referred to as a sampled trajectory point. Hence, a candidate path is a smooth spline connecting the sampled trajectory points along the *s* direction. With the segmentation and trajectory point sampling in Figure 6.6, there are 16 possible trajectory points (4 in the *s* direction, and 4 in the *l* direction), but we need only consider the splines which connect trajectory points along the increasing *s* direction since backward driving is not acceptable in normal urban road driving conditions (it will be dealt with separately, as a special case, though). The total number of candidate paths is 4^4 = 256, among which the path planning problem is to search for an optimal path with minimal cost while respecting certain constraints.

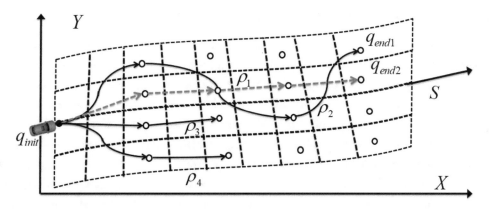

Figure 6.6: Possible candidates in path planning with divided road grids and sampling points in a lane-based SL-coordinate system.

We use polynomial spirals to connect sampled trajectory points. Polynomial spirals represent a cluster of curves whose curvature could be represented as polynomial functions of the arc length (along the s direction). The degree of polynomial spiral is not essential, and we use cubit or quintic spirals, whose arc length s and curvature κ satisfy:

$$\kappa(s) = \kappa_0 + \kappa_1 s + \kappa_2 s^2 + \kappa_3 s^3 \text{ or } \kappa(s) = \kappa_0 + \kappa_1 s + \kappa_2 s^2 + \kappa_3 s^3 + \kappa_4 s^4 + \kappa_5 s^5.$$

The only significant difference between cubic and quintic spirals is in satisfying boundary constraints: the second-order derivative of curvature $d\kappa^2 / ds^2$, which corresponds to the steering wheel rotating speed, is not continuous in cubic spirals $d\kappa / ds$ and $d\kappa^2 / ds^2$ continuous. Note that, when speed is low, the discontinuity introduced by cubic spirals is not a significant for downstream feedback control. However, such discontinuity cannot be safely ignored at high speeds.

The parameters of the proposed polynomial spirals connecting sampled trajectory points can be effectively obtained by a gradient descent fashioned search algorithm. For example, consider cubic spirals ($\kappa(s) = \kappa_0 + \kappa_1 s + \kappa_2 s^2 + \kappa_3 s^3$) connecting the initial pose $q_{init} = (x_I + x_I + \theta_I + \kappa_I)$ to the destination pose $q_{goal} = (x_G + x_G + \theta_G + \kappa_G)$ with continuous curvature. At the start s of the path, the first- and second-order derivative of curvature need to satisfy the following initial constraints:

$$\kappa_0 = \kappa_1$$
$$\kappa_1 = d\kappa(0) / ds$$
$$\kappa_2 = d^2\kappa(0) / ds^2.$$

This makes the actual unknown parameters to be the pair (κ_3, s_G), which could be quickly found by gradient descent search.

Find the Minimum Cost Path by Dynamic Programming

As we described, path planning can be described as a search to find the minimum cost path among the $|l_{total}/\Delta|^{|s_{total}/\Delta s|}$ candidate paths connecting the trajectory points along s-direction among $|l_{total}/\Delta l|*|s_{total}/\Delta s$ sampled trajectory points. Assuming that these sampled trajectory points constitute a graph $G = (V,E)$, where each trajectory point is a node in this graph: $v \in V, v = (x,y,s,l)$. For any two points $v, u \in V$, if their s coordinates satisfy $s_v < s_u$, we use $e(v,u) \in E$ to denote the cubic or quintic spiral from v to u. The optimal path-planning problem could be further converted into the problem of searching for a lowest-cost path (shortest path) on a directed weighted graph. What is special here is that the shortest path not only includes costs accumulated along the currently ex-panded path, but could also incorporate a potential cost associated with the newly expanded path if the expansion of a node is established. Consider the path τ connecting $n_0, n_1, \dots n_k$, where the initial point is n_0 and the end point is n_k, the cost of this established trajectory could be written as:

$$\Omega(\tau) = c(\tau) + \Phi(\tau),$$

where $c(\tau)$ represents the cumulated cost of following the path. Furthermore, $\Phi(\tau)$ is the cost intro-duced by ending the planned path at the end point n_k. If the $\Phi(\tau)$ function is the incremental cost introduced by ending at n_k, this means that:

$$\Omega(\tau(n_0, n_1, \dots n_k)) = g(n_k) + \Phi_c(\tau(n_{k-1}, n_k)),$$

where $g(n)$ is the minimum cost for reaching the node n. Note that cost incurred by following the spiral does not include the additional cost introduced by ending the path at n. Consider all the paths with n_{k-1} as their penultimate trajectory point, we would need to find the last trajectory point n_k which minimizes the total cost. More specifically, the node n_k satisfies the following properties.

1. There exists a directed edge $e(n_{k-1}, n_k)$, (denoted $\tau(n_{k-1}, n_k)$ which connects n_{k-1} and n_k.

2. For the set of nodes which n_{k-1} could reach (an edge $e(n_{k-1}, \tilde{n}_k)$ exists) (denoted $\{\tilde{n}_k\}$), the trajectory which ends at n_k has the lowest total cost: $n_k \leftarrow \underset{\tilde{n}_k}{\operatorname{argmin}} \; g(n_{k-1}) + c(\tau(n_{k-1}, \tilde{n}_k)) + \Phi_c(\tau(n_{k-1}, \tilde{n}_k))$ where c is the cost of the spiral connecting trajectory points $n_{k-1}, \tilde{n}_k$.

Hence, we could update $g(n_{k-1})$ as: $g(n_k) \leftarrow g(n_{k-1}) + c(e(n_{k-1}, n_k))$.

We could use the Dynamic Programming algorithm shown in Figure 6.7 to compute the optimal path with minimum cost, starting from the initial node n_0, connecting the trajectory points along increasing longitudinal s-direction. Note that, in the algorithm, connections between two trajectory points are computed in an ad-hoc fashion while searching for the optimal path on the graph. The $g(n)$ represents the cost of merely reaching the node n, and $\phi(n)$ represents the current path cost to node n, which includes both the cost of merely reaching node n and the additional cost of ending the path with node n. The former term $g(n)$ measures the additional cost incurred while choosing nodes to expand from the current node to the successor nodes (Figure 6.7, line 13). The latter term $\phi(n)$ is the criteria used when considering which predecessor nodes expand to current node (Figure 6.7, line 11). When all the computations of $g(n)$ and $\phi(n)$ have completed, it is easy to traverse the predecessor node map $prev\_node$ to construct the trajectory point with optimal (minimum) cost. Since candidate paths end at different trajectory points, we create a virtual node n_f and construct virtual edges connecting the last trajectory points to n_f. Our task then becomes finding a path connecting node n_f to node n_f with minimum cost. As shown in the algorithm, after computing $g(n)$ and $\phi(n)$, the last actual trajectory point can be found by the algorithm in Figure 6.7.

```
1 function Search_DP(TrajectoryPointMatrix(V,E), {s}, {l})
2     Initialize map g : ∀n∈V, g(n)← inf
3     Initialize map prev_node : ∀n∈V, g(n)← null
4     for each sampled sᵢ ∈ {s} :
5         ∀n∈V s.t. s(n)=sᵢ : φ(n)← inf
6         for each lateral direction Trajectory Point n =[sᵢ,lⱼ]:
7             if g(n)≠inf :
8                 Form the vehicle pose vector x̂ₙ =[x(n),y(n),θ(n),κ(n)]
9                 for each outgoing edge ẽ =(n,n')
10                    Form the polynominal spiral τ(ẽ(n,n'))
11                    if g(n)+Φ_C(n)<φ(n'):
12                        φ(n')← g(n)+Φ_C(n)
13                        g(n')← g(n)+c(τ)
14                        prev_node(n')← n
15                    end if
16                end for
17            end if
18        end for
19    end for
```

Figure 6.7: Finding the minimum cost path connecting trajectory points with Dynamic Programming.

The cost function in Figure 6.7, line 13, would require certain design considerations. Consider the trajectory cost $\Omega(\tau) = c(\tau) + \Phi(\tau)$, where $c(\tau)$ represents the cost of the spiral connecting two trajectory points adjacent in the s-direction; the following factors need to be taken into consideration while designing the cost function.

- **Road-map related aspect:** We want the planned spiral paths to be close to the central reference line of lanes. For example, when the behavioral decision is to follow a straight lane, the planned path will have a larger cost when the planned path shows a larger lateral distance (not approaching the central reference line).

- **Obstacle related aspect:** The planned path will have to be collision free. For example, with the grid division on the SL-coordinate lanes (Figure 6.6), any grid that is occupied by any obstacle, along with their proximate grids, should be assigned with extremely high costs to guarantee safety. Note that the "collision freeness" is mostly about static obstacles in path planning, which is only in the spatial (physical) space. How to guarantee safety, especially for avoiding non-static obstacles in the spatial-temporal dimensions, will be addressed in the discussion on speed planning.

- **Comfortableness and control feasibility:** The shape of the planned spiral paths should be smooth. This usually means smooth curvature change and slow change of derivatives of the curvature. In addition, not only the individual slices of planned spiral paths should be smooth, but also the connections between two spirals.

As for the $\Phi(\tau)$ cost, which is more about the trajectory in general, we could only consider the longitudinal s-distance since speed planning will further address the problem. One way to design the cost function $\Phi(\tau)$ is from [9]:

$$\Phi(\tau) = -\alpha s_f(\tau) + h_d(s_f(\tau)),$$

$$h_d(s) = \begin{cases} -\beta \text{ if } s \geq s_{\text{threshold}} \\ 0 \quad \text{otherwise.} \end{cases}$$

The first term $-\alpha s_f(\tau)$ is the linear cost that gives preference to longer s by giving a discount, and the second term is a nonlinear cost which only gets triggered if s is larger than a threshold. In this context, the alpha is a coefficient for this term. This coefficient is addressed as a "discount."

Speed Planning with ST-Graph

After a path has been determined by the path planning, the motion planning module will compute how fast the autonomous vehicle will traverse along this path, which we refer as the *speed planning* problem. The inputs for speed planning are a few candidate paths, as well as the upstream behavior decisions. Note that the constraints for speed planning are usually imposed by physical limits and comfort concerns such as rate of acceleration and steering wheel direction changes. Since input paths are represented as sequences of points, the output of speed planning must associate these points with desired speed information such as velocity, acceleration, curvature, and even their higher-order derivatives. Static obstacle information such as road shape has already been taken into consideration while doing path planning. In this section, we will introduce the concept of *ST-graph* to show how we formalize the speed planning problem into a search optimization problem on the *ST-Graph*.

In a typical *ST-Graph* (see Figure 6.8), two dimensions are usually considered, where *T* is the time axis and the *S* axis represents the tangential or longitudinal distance traveled along a given path. It is important to remember that when we speak of an *ST-Graph*, there is always an underlying pre-determined path. In addition, the *ST-Graph* does not necessarily have to be 2D, but could also be a 3D *SLT-Graph*, where the additional dimension is the lateral distance perpendicular to the trajectory, denoted as the *L* dimension. In Figure 6.8, we use a detailed example to demonstrate how speed planning is done via a 2D ST-graph.

Consider a trajectory that represents a lane switch by trajectory planning, with two obstacle vehicles (**a** and **b**) at the destination lane (the left adjacent lane of the autonomous vehicle). Without loss of generality, consider that traffic prediction results in these two vehicles both following their current lane at a constant speed. These prediction results for a and b will have their perspective area covered as shown by the greyed areas in the *ST-graph* in Figure 6.8. At any moment, projections of vehicles **a** and **b** onto the *ST-graph* are always line segments parallel to the *S*-axis, while these *S*-axis parallel line segments are dragged along the *t*-axis as vehicles **a** and **b** move along the *ST-graph* trajectory, leading to the greyed area (greyed quadrilaterals) in Figure 6.8. Just like in path planning, we also divide the *ST-graph* area uniformly into smaller *Lattice Grid* and associate each grid with a cost. Then speed planning can be formalized as a minimal cost path search problem on the *ST-graph* lattice grids. At $t = 0$ our autonomous vehicle is at position $s = 0$, and it needs to reach $s = s_{end}$ eventually following a *ST-graph* path whose cost is minimal.

As shown in the figure, we compare three candidate speed planning paths on the *ST-graph*.

- **Speed Plane 1:** The first one (Speed Plan 1) represents a path always behind vehicles **a** and **b** at any moment. The slope of Speed Plan 1 represents the speed of our autonomous vehicle, and the path will eventually reach $s = s_{end}$ position (even though it is not explicitly drawn on the figure.) Speed Plan 1 means that our autonomous vehicle will *yield* both vehicle **a** and **b** by only entering the adjacent left lane after both **a** and **b** have passed.

- **Speed Plan 2:** The second trajectory obviously also starts from the same origin. However, the slope of the path keeps increasing until our autonomous vehicle reaches a certain speed. The corresponding *s*-direction distance will surpass vehicle **a** at some point, but never exceeds the *s* distance of vehicle **b** at any moment. In real-world execution, the speed planning result of Speed Plan 2 is a typical movement of *yield* leading vehicle **b** and *overtake* the trailing vehicle **a** by inserting the vehicle in the gap between **a** and **b**.

- **Speed Plan 3:** The autonomous vehicle accelerates until it passes both vehicle **a** and **b**. At all times, its s distance is always larger than vehicles **a** and **b**.

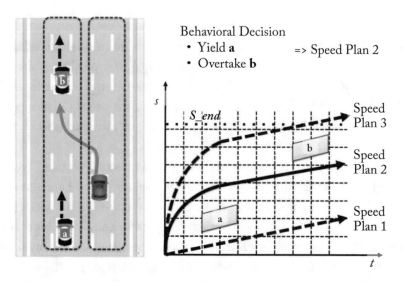

Figure 6.8: Speed planning via an *ST* graph.

Let's assume that the behavioral decision output for vehicle **a** is to "overtake" and to "yield" for **b**. Such behavioral decisions will assign lower costs to grids above the covered area of vehicle **a**, encouraging the algorithm to favor paths on the graph above **a**. Costs of grids below **b** will also be put lower to set algorithmic preferences over paths below **b**. Meanwhile, costs of grids below **a** or above **b** will get higher such that these areas could be avoided by the search algorithm.

Grids that are close enough to any obstacle will have very high costs to guarantee freedom from collisions. In addition, the speed-planning curve in general will have costs associated with accelerations. For example, if the connection between two points on the speed planning curve is too steep, which represents a very large or even discontinuous acceleration, then the cost associated with the acceleration needed for these two speed-planning points may be set extremely high. In fact, a sharp speed increase will lead to control failure due to the large acceleration. Note that control failure means that, at the moment, the vehicle's actual pose (position and velocity) is deviating too much from the desired pose as defined in the motion planning trajectory. Sometimes it is important to adjust the locations of speed-planning points inside a chosen grid to optimize the whole speed planning trajectory cost. The speed-planning trajectory on the *ST-graph* could be computed via graph search algorithms like A* or Dijkstra, given the cost profile of each grid and the grid definitions. After we have computed the speed-planning curve on the *ST-graph*, we could easily retrieve speed (the slope) and acceleration (the derivative of the slope) for each trajectory point we want to output, hence completing the motion planning phase. With the path-planning and speed-planning module described above, we have now computed the behavior decisions based on our destination and surrounding environments into a concrete trajectory with spatial and temporal information. We uniformly sample points from this spatial-temporal trajectory and send these sampled trajectory points with speed, acceleration, curvature, etc. to the downstream feedback control module. The feedback control module will output the actual control signals to physically manipulate the vehicle.

It is worth mentioning here that the *ST-graph* is not only a very intuitive method in solving speed planning, it is also an important concept in simultaneous longitudinal and lateral planning method (to be described in the next subsection).

6.2.3   MOTION PLANNING WITH LONGITUDINAL PLANNING AND LATERAL PLANNING

Instead of doing path planning and then speed planning, [10] proposed the idea of doing longitudinal and lateral planning. The longitudinal and lateral dimensions naturally fit into the SL-coordinate system described in Section 6.3.1, where s corresponds to the longitudinal direction and l to the lateral direction. In both the longitudinal and lateral dimensions, the planning problem occurs in a space very similar to the speed-planning problem in the *ST-graph*. In fact, a simple but straightforward approach could be to leverage the *ST-graph* solution and perform two graph searches on the *s-t* dimension and *l-t* dimension. However, the *ST-graph* fashioned approach we discussed above is quite suitable for obstacle avoidance and obeying upstream behavioral decisions, but is not a perfect method for determining the optimal desired spatial-temporal trajectory. A desired motion planning trajectory should be smooth, which is usually represented mathematically in continuity of pose with its derivatives in a dimension (e.g., *position*, *speed*, and *acceleration*). Ease and comfort can be best measured by the *jerk* parameter, which is the rate of acceleration. Consider the jerk-optimal trajectory connecting a start state $P_0 = [p_0, \dot{p}_0, \ddot{p}_0]$ and an end state $P_1 = [p_1, \dot{p}_1, \ddot{p}_1]$, where the pose P could be either the longitudinal s-pose or lateral l-pose, in a time frame of $T := t_1 - t_0$. Call the integral of the squared *jerk*: $J_t(p_t) := \int^{t_1} \ddot{p}^2(\tau) d\tau$. To minimize any cost function of the following general form:

$$C = K_j J_t + K_t g(T) + K_p h(p_1)$$

(where g and h are arbitrary functions and $K_j, K_t, K_p > 0$), the optimal solution is a quintic polynomial. The proof from an optimal control's perspective is clear since the end point costs $g(T)$ and $h(p_1)$ do not change the Euler–Lagrange equation (a formal proof is illustrated in [10]). Intuitively, the cost function penalizes a high jerk integral J_t along the trajectory, while also considering the time T and end pose $h(p_1)$.

Lateral Planning

We start the 1D planning with the lateral l-dimension. Then the start state P_0 becomes $D_0 = [d_0, \dot{d}_0, \ddot{d}_0]$, and we set this start state according to the actual end state of the previous planning frame, such that continuities are maintained. The end state P_1 is selected out of a number of feasible candidate lateral offset d's with $\dot{d}_1 = \ddot{d}_1 = 0$ since we favor moving in a direction parallel to the central reference line (s-direction). The functions g and h are chosen with $g(T) = T$ and $h(d_1) = d_1^2$. We can see that slow convergence is penalized as well as any lateral difference with $d = 0$ at the end state. The optimal solution takes the form of a quintic polynomial: $l_{optimal}(t) = a_5 t^5 + a_4 t^4 + a_3 t^3 + a_2 t^2 + a_1 t + a_0$, which minimizes the cost function of:

$$C_l = K_j J_t(l(t)) + K_t T + K_l d_1^2.$$

The coefficients $a_5, a_4, a_3, a_2, a_1, a_0$ can be calculated with boundary conditions at $D_0 = [d_0, \dot{d}_0, \ddot{d}_0]$ and $D_0 = [d_1, \dot{d}_1, \ddot{d}_1]$. We can select a set of candidate d_i's and compute a set of best candidate 1D trajectories as $Solution_{i,j}$ from $[d_1, \dot{d}_1, \ddot{d}_1, T]_{i,j} = [d_1, 0, 0, T_j]$. Each candidate solution will have a cost. For each candidate, we check if this solution is consistent with the upstream behavioral decision and make this solution invalid if any violations/collisions are found. The remaining valid trajectories propose a candidate set of lateral dimension, to be utilized in computing the planned trajectory on the 2D dimension. Note that in this context, the lateral trajectories are represented $l(t)$ functions. Therefore, we call these trajectories in the lateral dimension.

The method described above works well for high-speed trajectories where longitudinal movement and lateral movement can be independently chosen. However, this assumption does not remain valid at low speeds. At extreme low speeds, many of the generated trajectories in the lateral dimension will probably be invalid because of the non-holonomic property of vehicle control. When speed is low, a different mode of lateral dimension motion planning should be triggered. In this mode, the lateral trajectory is dependent on the longitudinal trajectory, and is denoted as $l_{low\text{-}speed}(t) = l(s(t))$. Since the lateral movement is dependent on the longitudinal movement, we must treat the lateral movement as a function of longitudinal position s. With the cost function being in the same format but dependent on s, the arc length rather than t, we modify the cost function to:

$$C_l = K_j J_t(l(s)) + K_t S + K_l d_1^2,$$

where $S = s_1 - s_0$ and all the derivatives are taken with respect to s instead of t. The optimal solution to this s-based cost function is a quintic function over s, and could be computed with the selected T_j and end state d_i's.

Longitudinal Planning

Longitudinal dimensional planning is similar except that the end state we want to track sometimes depends on a moving target. In detail, The longitudinal movement is affected by certain targets. For example, we might want to reach a certain speed in certain time. Or just would like to follow a lead vehicle. These targets/goals will define the "end state". For example, cruising (reaching a certain speed) defines the end-state velocity and does not care about the end-state position. Also take the following lead vehicle case for example, the end state would be: at a future time t, we want to be at a position (say x meters behind the lead vehicle), and with a speed (the same as the lead vehicle speed). In the case of the following, the end-state is defined by a vector consisting of s, ds (or namely v), and dds (or namely a). Let us use s_{target} (t) as the trajectory we would like to track. The start state is $S_0 = [s_0, \dot{s}_0, \ddot{s}_0]$. The candidate trajectory set could be chosen with different Δs_i (either positive or negative) and T_j:

$$S_{i,j} = [s_1, \dot{s}_1, \ddot{s}_1, T_j] = [(s_{target}(T_j) + \Delta s_i), \dot{s}_{target}(T_j), \ddot{s}_{target}(T_j), T_j].$$

We will now discuss cost function settings under three different behavioral decision scenarios.

Following Another Vehicle

The desired longitudinal movement for the autonomous vehicle along the s direction to follow a vehicle ahead is to maintain a minimum distance as well as a time gap behind the vehicle ahead. Here we want to emphasis the importance of output from the traffic *prediction* module, since s_{target} (t) will be dependent on the predicted trajectory $s_{front-vehicle}$ (t) of the vehicle ahead at the planning time frame/cycle. The target movement becomes:

$$s_{target}(t) = s_{front-vehicle}(t) - D_{min} - \gamma \dot{s}_{front-vehicle}(t),$$

which implies at least a distance D_{min} and a constant time gap γ given the front vehicle speed $\dot{s}_{front-vehicle}$ (t) at the end time. Given the desired end state as $[s_{desire}, \dot{s}_{desire}, \ddot{s}_{desire}]$, the cost function hereby becomes:

$$C_s = K_j J_t + K_t T + K_p (s_1 - s_{desire})^2.$$

If we assume that the predicted trajectory has a constant acceleration: $\ddot{s}_{front-vehicle}(t) = \ddot{s}_{front-vehicle}(t_0)$, then the speed $\dot{s}_{front-vehicle}$ (t) and position $s_{front-vehicle}$ (t) could be integrated as:

$$\dot{s}_{front-vehicle}(t) + \ddot{s}_{front-vehicle}(t_0) + \ddot{s}_{front-vehicle}(t_0)(t - t_0)$$

$$s_{front-vehicle}(t) = s_{front-vehicle}(t_0) + \dot{s}_{front-vehicle}(t_0)(t - t_0) + \frac{1}{2}\ddot{s}_{front-vehicle}(t_0)(t - t_0)^2.$$

The end-state longitudinal direction speed is $\dot{s}_{target}(t) = \dot{s}_{front\text{-}vehicle}(t) - \gamma \ddot{s}_{front\text{-}vehicle}(t)$ and acceleration is $\ddot{s}_{target}(t) = \ddot{s}_{front\text{-}vehicle}(t_1)$. Sometimes the *following* does not have a specific object: in case of no leading vehicle or no vehicle to follow, the cost function here will be then irrelevant to any front-vehicle or lead vehicle. This means that there is no object to follow. Here the following is referred to the "cost associated with following a vehicle" in the cost function described here:

$$C_s = K_j J_t + K_t T + K_p (\dot{s}_1 - \dot{s}_{desire})^2.$$

In this case of no object to follow but keeping the speed up, the optimal trajectory set will be quartic polynomials instead of quintic with the set of $\Delta \dot{s}_i$ and T_j. In detail, when there is no "front vehicle" or lead vehicle to follow, we only care about reaching a certain speed at a certain time.

Switching Lane by Yielding and/or Overtaking

Switching to an adjacent lane when yielding to a vehicle is similar to the previous case of following a vehicle in the longitudinal sense. Let's denote the target longitudinal movement to be: $s_{target}(t)$.

- When switching lanes by yielding to a vehicle **b**, the target trajectory becomes:
$$s_{target}(t) = s_b(t) - D_{min\text{-}yield} - \gamma \dot{s}_b(t).$$

- When switching lanes by overtaking a vehicle **a**, the target trajectory becomes:
$$s_{target}(t) = s_a(t) + D_{min\text{-}overtake} - \gamma \dot{s}_a(t).$$

- When merging between two vehicles **a** and **b** (for example, Figure 6.8), the target trajectory becomes:
$$s_{target}(t) = \tfrac{1}{2}[s_a(t) + s_b(t)].$$

Stopping

When the autonomous vehicle needs to stop at a pedestrian cross or traffic signal line, the same form of cost function can be used, and the target movement will become constant for the s-direction $s_{target}(t) = s_{stop}$ with first- and second-order derivatives being 0 ($\dot{s}_{target}(t) = 0$ and $\ddot{s}_{target}(t) = 0$).

After both longitudinal and lateral candidate trajectory sets have been computed, the optimal trajectory comes from $|Traj_{longitudinal}| \times |Traj_{lateral}|$ possible combinations. Each combined trajectory is checked against behavioral decision output for abeyance as well as collision free. In addition, trajectories which push toward feedback control limit will also be filtered out. The final trajectory will be selected from the remaining valid trajectories with minimum weighted longitudinal and lateral cost.

6.3   FEEDBACK CONTROL

From a stand-alone controlling point of view, the feedback control module of autonomous driving has no essential difference with general mechanical control. Indeed, both autonomous vehicle control and general mechanical control are based on certain pre-defined trajectories and track the difference between the actual pose and the pose on the pre-defined trajectory by continuous feedback. [11] lists many different works on autonomous vehicle feedback control. For instance, [8, 12] added additional obstacle avoidance and route optimization to the traditional feedback control system. With our proposed system architecture of autonomous planning and control, the feedback control module in our system will mostly leverage existing work of traditional vehicle pose feedback control. Since this part of the work is relatively mature and well understood, it is not the focus of this autonomous driving book. However, for some background on feedback control for autonomous driving, we will just discuss two important concepts: the Vehicle Bicycle Model and the PID Feedback Control System [13, 14]. A more detailed description of other feedback control systems in autonomous driving can be found in [12].

6.3.1   BICYCLE MODEL

In Section 3.1, we briefly introduced a vehicle model to better describe the trajectory generation algorithm in Motion Planning. Here we will describe in more detail a frequently used vehicle model in autonomous driving Feedback Control: the *Bicycle Model*. The pose represented by the bicycle model is within a 2D plane. The vehicle pose can be fully described by the central position (x, y) and the heading angle θ between the vehicle and the 2D plane's x-axis. Under this model, the vehicle is considered to be a rigid body with the front and rear wheels connected by a rigid axis. The front wheels can freely rotate within a certain angle range, while the rear wheels remain parallel to the vehicle body and cannot rotate. The rotation of the front wheels corresponds to the position of the steering wheel. An important characteristic of the bicycle model is that vehicles cannot make lateral movements without moving forward (making longitudinal movements), and such a characteristic is also addressed as a non-holonomic constraint. Under this vehicle model, the non-holonomic constraint is usually expressed as differential derivative equations or inequations. We also neglect the inertial and slippery effect at the contact point between the tire and the ground surface. Note that this assumption does not have any significant impact at low speeds and brings very little error. However, at high speeds, the effect of inertial on feedback control is significant and cannot be safely ignored. The physical vehicle model at high speeds with inertial effect is more complex [12] and will not be explored further within the context of this book.

The vehicle pose representation in the bicycle model is shown in Figure 6.9. We use an xy-based plane as the 2D plane, where $\hat{e}_x$ and $\hat{e}_y$ separately represent the unit vector in the x and y directions. p_r and p_f stand for the front-wheel contact point and the rear-wheel contact point. The heading angle θ is the angle between the vehicle and the x-axis (angle between vector p_r and unit vector $\hat{e}_x$). The steering wheel rotation angle δ is defined as the angle between the front wheel direction and the vehicle body, where the ground contact points of front and rear wheels (p_f and p_r) satisfy the following properties:

$$(\dot{p}_r \cdot \hat{e}_y)\cos(\theta) - (\dot{p}_r \cdot \hat{e}_x) = 0$$
$$(\dot{p}_f \cdot \hat{e}_y)\cos(\theta + \delta) - (\dot{p}_f \cdot \hat{e}_x)\sin(\theta + \delta) = 0,$$

where $\dot{p}_f$ and $\dot{p}_r$ are the instant speed vector of the front and rear wheel at their ground contact point. Consider the scalar projections of the rear wheel speed at the x-axis and y-axis: $x_r := p_r \cdot \hat{e}_x$ and $x_y := p_r \cdot \hat{e}_y$, along with the tangential speed at the rear wheel $v_r := \dot{p}_r \cdot (p_f - p_r)/\| p_f - p_r \|$, then the above constraints between p_f and p_r can also be written as:

$$\dot{x}_r = v_r \cos(\theta)$$
$$\dot{y}_r = v_r \sin(\theta)$$
$$\theta = v_r \tan(\delta)/l,$$

where l represents the length of the vehicle (distance between the front axis center and the rear axis center). Smilarly, the relationship between the front wheel variables can be written as:

$$\dot{x}_f = v_r \cos(\theta + \delta)$$
$$\dot{y}_f = v_r \sin(\theta + \delta)$$
$$\theta = v_f \sin(\delta)/l.$$

Note that the scalar variables of front and rear wheel speeds satisfy: $v_r = v_f \cos(\delta)$.

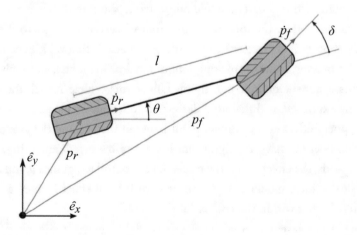

Figure 6.9: Bicycle model in feedback control [5], used with permission.

With the bicycle model described above, controlling means finding the steering wheel $\delta \in [\delta_{min}, \delta_{max}]$ and forward speed $v_r \in [\delta_{min}, \delta_{max}]$ parameters which satisfy the physical pose constraints. In practice and for simplicity, the control outputs are the steering-wheel angle change rate ω and the throttle/brake percentage instead of actual steering wheel or forward speed goals. The relationship between ω and δ is simplified as: $\tan(\delta)/l = \omega/v_r = \kappa$. This means that the problem is reduced to finding the steering wheel change rate δ that satisfies the constraints. Such simplification is called the *Unicycle Model*, which has the characteristic that the forward speed has been simplified to be only dependent on the vehicle axis length and the steering angle change rate.

6.3.2 PID CONTROL

The most typical and widely used algorithm in autonomous vehicle feedback-control is the PID feedback control system, as shown in Figure 6.10, where the term $e(t)$ represents the current *tracking error* between desired pose variable and actual pose variable. The variable to track could be the longitudinal/lateral difference along a trajectory, the angle/curvature difference at various trajectory points, or even a comprehensive combination of these vehicle pose variables. In Figure 6.10, the P controller represents the feedback for the current tracking error, whose coefficient is governed by K_p ; the I and D controllers represent the integral and differential part, whose coefficients are separately governed by K_I and K_D.

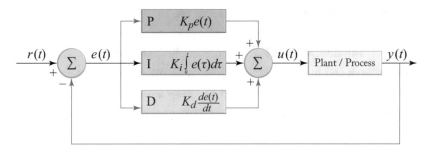

Figure 6.10: PID-based feedback-control system (based on [13]).

As for the feedback control module in autonomous vehicle, the task is to control the vehicle to follow the upstream motion planning output trajectory as closely as possible. We thus propose using the methodology in [15], and leveraging two PIC controllers to separately control the steering-wheel angle δ and the forward speed V_s. At a given time frame n, the PID controller for the steering-wheel angle is as follows:

$$\delta_n = K_1\theta_e + K_2 l_e / V_s + K_3 \dot{l}_e + K_4 \sum_{i=1}^{n} l_e \Delta t .$$

The variables θ_e and l_e are all tracking error terms between the actual pose and the desire pose on the motion-planning output trajectory point at this time frame n. For each time frame at this time frame n, the corresponding pose on the motion planning output trajectory point is addressed as the *reference point*. θ_e represents the angle difference between the vehicle pose heading and the reference point heading, while l_e tracks lateral difference between the vehicle actual lateral position and the reference point lateral position. V_s is the forward speed. The coefficient K_1 and K_2 are meant for the P controller, while K_3 governs the differential part (D controller) and K_4 the integral part (I controller). Given that this steering-angle controller serves for direction, the other PID controller is more about the forward speed V_s along the longitudinal direction (*s*-direction), and controls throttle/brake output. This controller considers the difference between the actual vehicle pose curvature and the reference point curvature. From these curvatures, we can design a function to track the forward speed error. Given this desired forward speed to track and the actual forward speed xxx, the PID controller for the forward speed can be written as:

$$V_e = V_{desired} - V_s$$
$$U_V = K_p V_e + K_I \sum V_e \Delta t + K_D \Delta V_e / \Delta t,$$

where K_p, K_I, and K_D separately represent the gain for the proportional, integral, and differential part, and U_V represents throttle/brake output for this given time frame n.

These two PID controllers discussed here are the most typical and basic implementation practices for the feedback *control* module in autonomous vehicles. To make an even better passenger experience in autonomous driving, more complicated feedback control systems will be necessary to further track and tune variables such as curvature and jerk. The problem of generating delicate and accurate control to enforce an object to follow a pre-defined trajectory is not a unique problem for autonomous driving, there are many existing solutions [15].

6.4   CONCLUSIONS

With Chapters 5 and 6, we have comprehensively described the general architecture of autonomous vehicle planning and control, consisting of *routing*, behavioral *decision*, traffic *prediction*, motion *planning*, and feedback *control*. All these modules have existing and working proposals in both academia and industry. These existing proposals exhibit different strengths, either in solid theoretical foundation or practical industry practices. In fact, we believe the challenge of autonomous driving does not lie in solving the problem at any individual level or module, but rather in the philosophy behind how we partition the large autonomous vehicle planning and control problem into distinctive layers and solve these smaller problems in a coherent fashion. From this perspective, we do not aim at bringing all possible solutions in a survey-like fashion. However, we do bring to the reader a consistent series of working solutions for each module. Our focus has been to illustrate clearly what is the problem scope and definition for each module, and how together they solve the general problem following the data flow from abstractness to concreteness. In each module or layer, the problem is formalized and solved with practical industrial solutions. We also discussed how each downstream layer will utilize the upstream module output as input and further compute more concrete solutions toward the data flow to actual control signals.

6.5   REFERENCES

[1]   Montemerlo, M., Becker, J., Bhat, S., Dahlkamp, H., Dolgov, D., Ettinger, S., Haehnel, D., Hilden, T., Hoffmann, G., Huhnke, B., Johnston, D., Klumpp, S., Langer, D., Levandowski, A., Levinson, J., Marcil, J., Orenstein, D., Paefgen, J., Penny, I., Petrovskaya, A.,Pflueger, M., Stanek, G., Stavens, D., Vogt, A., and Thrun, S. 2008. Junior: The Stanford entry in the urban challenge. *Journal of Field Robotics: Special Issue on the 2007 DARPA Urban Challenge*, 25(9), pp. 569–597. 112

[2]   Urmson,C. , Anhalt, J., Bagnell, D., Baker, C., Bittner, R., Clark, M. N., Dolan, J., Duggins, D., Galatali, T., Geyer, C., Gittleman, M., Harbaugh, S., Hebert, M., Howard,T. M., Kolski, S., Kelly, A., Likhachev, M. McNaughton, M., Miller, N., Peterson, K., Pilnick, B., Rajkumar, R., Rybski, P., Salesky, B., S, Y. W., Singh, S., Snider, J., Stentz, A., Whittaker, W., Wolkowicki, Z., Ziglar, J., Bae, H., Brown, T., Demitrish, D., Litkouhi, B., Nickolaou, J., Sadekar, V., Zhang, W., Struble, J., Taylor, M., Darms, M., and Ferguson, D. 2008. Autonomous driving in urban environments: Boss and the urban challenge. *Journal of Field Robotics: Special Issue on the 2007 DARPA Urban Challenge*, 25(9), pp. 425–466. 112

[3]   Buehler, M., Lagnemma, K., and Singh, S. (Eds.) 2009. *The DARPA Urban Challenge: Autonomous Vehicles in City Traffic*. Springer-Verlag Berlin Heidelberg. DOI: 10.1007/978-3-642-03991-1. 112

[4]   Katrakazas, C., Quddus, M., Chen, W. H., and Deka, L. 2015. Real-time motion planning methods for autonomous on-road driving: State-of-the-art and future research directions. Elsevier Transportation Research Park C: *Emerging Technologies*, Vol. 60, pp. 416–442. DOI: 10.1016/j.trc.2015.09.011. 112, 123

[5]   Paden, B., Cap, M., Yong, S. Z., Yershow, D., and E. Frazzolo. 2016. A survey of motion planning and control techniques for self-driving urban vehicles. *IEEE Transactions on Intelligent Vehicles*, 1(1), pp. 33–55. DOI: 10.1109/TIV.2016.2578706. 112, 123, 138

[6]   Brechtel,S. and Dillmann, R. 2011. Probabilistic MDP-behavior planning for cars. *IEEE Conference on Intelligent Transportation Systems*. DOI: 10.1109/ITSC.2011.6082928. 114, 115

[7]   Ulbrich, S. and Maurer, M. 2013. Probabilistic online POMDP decision making for lane changes in fully automated driving. *16th International Conference on Intelligent Transportation Systems (ITSC)*. DOI: 10.1109/ITSC.2013.6728533. 114, 115, 123

[8]   Gu, T., Snider, J., M. Dolan, J. and Lee J. 2013. Focused trajectory planning for autonomous on-road driving. *IEEE Intelligent Vehicles Symposium (IV)*. DOI: 10.1109/IVS.2013.6629524. 123, 137

[9]   Mcnaughton, M. 2011. Parallel algorithms for real-time motion planning. Doctoral Dissertation. Robotics Institute, Carnegie Mellon University. 123, 124, 126, 130

[10]   Werling, M., Ziegler, J., Kammel, S., and Thrun, S. 2010. Optimal trajectory generation for dynamic street scenarios in a frenet frame. In *2010 IEEE International Conference on Robotics and Automation (ICRA)*, (pp. 987–993). IEEE. DOI: 10.1109/ROBOT.2010.5509799. 133

[11]   Zakaria, M. A., Zamzuri, H., and Mazlan, S. A. 2016. Dynamic curvature steering control for autonomous vehicle: Performance analysis. *IOP Conference series: Materials Science and Engineering, 114,* 012149. DOI: 10.1088/1757-899X/114/1/012149. 137

[12]   Connors, J. and Elkaim, G.H. 2008. Trajectory generation and control methodology for an ground autonomous vehicle. *AIAA Guidance, Navigation and Control Conference.* 132, 137

[13]   https://en.wikipedia.org/wiki/PID_controller. 137, 139

[14]   National Instruments: http://www.ni.com/white-paper/3782/en/. 137

[15]   Zakaria, M.A., Zamzuri, H., and Mazlan, S.A. 2016. Dynamic curvature steering control for autonomous vehicle: Performance analysis. *IOP Conference series: Materials Science and Engineering, 114,* 012149. DOI: 10.1088/1757-899X/114/1/012149. 139, 140

CHAPTER 7

Reinforcement Learning-Based Planning and Control

Abstract

While optimization-based approaches still enjoy mainstream appeal in solving motion planning and control problems, learning-based approaches have become increasingly popular with recent developments in artificial intelligence. Even though current state-of-the-art learning-based approaches to planning and control have their limitations, we feel they will become extremely important in the future and that, as technical trends, they should not be overlooked. More particularly, reinforcement learning has been widely used in solving problems that take place in the form of rounds or time steps with step-wise guiding information such as rewards. Therefore, it has been experimented with as a methodology to solve different levels of autonomous driving planning and control problems. We thus conclude that reinforcement learning-based planning and control will gradually become a viable solution to autonomous driving planning and control problems or at least become a necessary complement to current optimization approaches.

7.1 INTRODUCTION

In the previous two chapters, we described the planning and control framework, which in a general sense includes modules of routing, traffic prediction, behavioral decision, motion planning, and feedback control. Our proposed behavioral decision solution uses a hierarchy of scenarios and rules to ensure safety. In both the motion planning and the feedback control modules, we are actually solving two optimization problems under certain distinct constraints. While this traditional approach of planning and control is the current mainstream approach, learning-based approaches [1, 2, 3, 8, 9] have emerged and have attracted increased interest from researchers. In practice, these optimization-based approaches have been working reasonably well in practice [5, 6].

However, given the current success of optimization-based planning and control approaches, we need to pay some attention to learning-based solutions. Especially, we would like to put our emphasis on reinforcement learning-based planning and control approaches in this chapter, for which there are three reasons. First, we believe that autonomous driving is still in an early stage and the current application scenarios are not as challenging as driving in a real unrestricted urban environment. Most car makers and autonomous driving technology companies test their autonomous driving vehicle in a road-testing fashion, usually in limited areas or even certain restricted routes. As of today, no autonomous driving vehicle products can meet the actual requirement of L4 [10] autonomous driving level in a relative large and unrestricted urban area. Under the limited scenarios, such as closed-environment highway or closed-environment park (airport, factory area etc), it is impossible to guarantee the sufficiency of optimization approaches to tackle all possible real-world unrestricted road cases.

The second argument against a purely optimization-based approach is that historical driving data has not been fully utilized. In detail, optimization approaches are usually based on formulating the planning problem into a mathematical optimization where the goal is to maximize certain rewards (minimize certain costs) given certain constraints. Such optimization goals and constraints are usually based on human experience or heuristics. That being said, representing human experience and/or heuristics is limited and it is hard to cover all the corner cases. Furthermore, human drive data are not being explicitly used. If human driving historical data can be leveraged in some form, it might help cover corner cases or complicated cases where simply stacking implementations of rules is insufficient. This is particularly true in the age of big data where data is becoming critically important as a utility and large amounts of driving data, either from human drivers or autonomous vehicles, have been accumulated. How to leverage such big data of historical driving to enhance planning and control of autonomous vehicles is still a challenge. It is obvious that valuable information can be mined and learned from such accumulated driving data, but this has not yet been successfully done. Traditional optimization-based approaches have very few ways to fully utilize these databases. However, learning-based methods could naturally make use of historical driving data. In addition to these two arguments, the last but most important argument is that a human driver learns how to drive, and in most cases from a teacher or coach rather than optimizing a cost or goal. This is the most significant argument for the adoption of reinforcement learning-based planning and control approaches: the reinforcement learning process is done through iteratively interacting with the environment through actions, which is very similar to how human drivers learn to drive from a coach's feedback and doctrine. In this sense, reinforcement learning has already been used in robotics control and we do believe that it will also play a critical role in the future of autonomous driving planning and control.

In this chapter, we will first introduce the basics of reinforcement learning. Then we will discuss two popular algorithms in reinforcement learning, Q-learning, and Actor-Critic learning methods. Reinforcement learning practices on autonomous driving will also be discussed. These reinforcement learning-based solutions land on various layers of autonomous driving planning and control. Some model the behavioral level decisions via reinforcement learning [3, 9], while other approaches directly work on the output of the motion planning trajectory or even direct feedback control signals [2]. In addition, reinforcement learning has demonstrated success under a range of distinct scenarios [1]. Nevertheless, general autonomous vehicle planning and control under unrestricted urban scenarios still remains a very challenging and not fully solved problem for reinforcement learning.

7.2 REINFORCEMENT LEARNING

The key characteristic of reinforcement learning is that the learning process is interactive with an *environment* and making reinforcement learning a close-loop learning process, as shown in Figure 7.1. The main entity for reinforcement learning is called the agent, which makes decisions by taking actions. Everything outside the agent is denoted as the environment. The reinforcement learning is a process in which the agent interacts with the environment by taking actions, sensing states, and getting rewards in an iterative fashion.

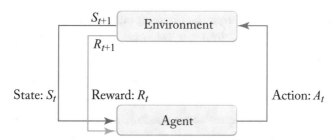

Figure 7.1: Reinforcement learning framework: agent interacting with environment by taking actions, sensing state, and getting rewards.

More specifically, the learning process takes place in rounds indexed by time $t = 0,1,2,3\ldots$. At each time step t, the agent perceives the environment by receiving the state of environment, denoted as $S_t \in S$, where S is the state space. The agent then must make a decision by taking an action $A_t \in A(S_t)$, where $A(S_t)$ is the action space given the state observation S_t. After the action is taken and at time $t + 1$, the environment interactively changes to state S_{t+1} and gives a scalar reward denoted R_t. The updated state S_{t+1} is immediately perceived by the agent and the reward R_t is also instantly received by the agent. How the reward is generated by the environment is called the *reward function*. A reward function defines the instant reward, which maps state (or state-action pair) of the environment to a scalar indicating the immediate intrinsic desirability of such a transition. The goal of the agent is to learn how to take actions so that the total cumulated rewards along the whole process will be maximized, as opposed to the reward function, which represents the instance desirability of state transition by taking an action, the total cumulated reward is called the *returns*. Assuming that the rewards after time step t are $R_{t+1}, R_{t+2}, R_{t+3}\ldots$, one simple form of return we could seek to maximize is simply the sum of rewards:

$$G_t = R_{t+1}, R_{t+2}, R_{t+3}\ldots + R_T,$$

where T is the last time-step. This kind of return to maximize is suitable for scenarios where the agent interacts with the environment in limited finite steps (e.g., two players playing poker). Here each valid sequence of states starts from certain states and always ends at a set of states called the *terminal states*. Each valid sequence of states is called an *episode*, and tasks of this kind are *episodic learning tasks*.

However, there are also reinforcement learning problems where a valid sequence of states might last forever. Tasks of this kind are called *continuing tasks*. In such cases, it is inappropriate to just sum up the rewards since the final time step is infinite. Instead, the return function, or the goal to maximize, becomes the expected discounted rewards:

$$G_t = R_{t+1} + \gamma R_{t+2} + \gamma^2 R_{t+3}\ldots = \sum_{k=0}^{\infty} \gamma^k R_{t+k+1},$$

where $0 \leq \gamma \leq 1$ is a parameter indicating the discount rate of how we value a future reward projected to one time-step more current. In detail, this parameter "gamma" discounts future reward. For example, reward amount X at time t is valued as gamma*X at time t-1. This is what we mean by "projected to one time-step more current." It is mathematically trivial to show that if the reward at any time-step is bounded, then the cumulated discounted rewards will also be bounded. In the extreme case of $\gamma = 0$, the agent only cares about maximizing immediate rewards.

At each time step, the agent needs to choose an action to take given the state perceived. The agent must follow a strategy that will maximize the expected returns. Such a strategy is also called a *policy*, and is denoted as π. A policy π defines the strategy of how the learning agent behaves at any given state S_t, and it could take forms as simple as a state-action look-up table, or as complex as a DNN. We will describe the detailed computational approximation of an optimal policy in the following sections. Formally, the policy π is a function from $\pi{:}S{\rightarrow}A$, which maps any given state $S_t \in S$ at time t to an action where $A_t \in (AS_t)$, and the probability of action a being chosen at state s is denoted as $\pi_t (a \mid s)$. The way a policy picks an action is usually associated with another function $V_\pi (s)$, called the *value function*. The value function measures how good or bad it is for an agent to be in a given state s. Formally, the value function measures the expected return for entering a state s and sticking to this policy afterward. For a simple Markov Decision Process (MDP, as described in Chapter 6) where the state transitions are Markov, the value function could be written as: $V_\pi(s) = E_\pi (G_t \mid S_t = s) = E_\pi (\sum_{k=0}^{\infty} \gamma^k R_{t+k+1} \mid S_t = s)$, where E_π is the expectation operator given the policy π. This value function is also called the *state–value function for policy π*. Similarly, the Q-value function is the mapping of state-action pairs to a scalar value representing the expected return at state S_t after taking action A_t and following the policy π afterward. It is denoted as: $Q_\pi (s,a) = E_\pi (G_t \mid S_t = s, A_t = a) = E_\pi (\sum_{k=0}^{\infty} \gamma^k R_{t+k+1} \mid S_t = s, A_t = a)$.

For many algorithms, the process of reinforcement learning is mostly estimating such state value function or state-action value functions.

A very unique property of the way value functions Q_π and V_π are defined is its inherent recursive structure. Such recursive structures have particularly elegant formalizations when the learning environment is Markovian. One episode in this MDP will be represented as the sequence:

$$s_0, a_0, r_1, s_1, a_1, r_2, s_2, a_2, r_3, s_3, \ldots s_{T-1}, a_{T-1}, r_T, s_T .$$

In an MDP process, given any policy π and state s, the value function of $V_\pi(s) = E_\pi(G_t \mid S_t = s)$ could be expanded as:

$$V_\pi(s) = E_\pi(G_t \mid S_t = s) = \sum_a \pi(a|s) \sum_{s'} p(s' \mid s, a)[r(s, a, s') + \gamma V_\pi(s')].$$

This is called the *Bellman Equation* for the value function $V_\pi(s)$. [7] has offered a thorough description of reinforcement learning algorithms.

7.2.1   Q-LEARNING

Q-Learning [7] is one of the most popular algorithms used for reinforcement learning. The idea is to learn to approximate the strategy that maximizes the expected return at any given state s_t by taking action a_t and then following the optimal strategy: $Q(s_t, a_t) = \max_\pi R_{t+1}$. How we then select the policy is based on: $\pi(s_t) = \mathrm{argmax}_{a_t} Q(s_t, a_t)$. The key problem in Q-learning is how to accurately estimate the Q-function that maps state-action pairs to an optimal expected return.

The Q-function $Q(s_t, a_t)$ exhibits the following recursive structure: $Q(s_t, a_t) = r(s_t, a_t, s_{t+1}) + \gamma \max_{a_{t+1}} Q(s_{t+1}, a_{t+1})$. This is the Bellman Equation in terms of the Q-function. Based on this recursive representation of $Q(s_t, a_t)$, the Q-function could be solved by Dynamic Programming in a backward propagation fashion:

$$Q(s_t, a_t) \leftarrow Q(s_t, a_t) + \alpha[r(s_t, a_t, s_{t+1}) + \gamma \max_{a_{t+1}} Q(s_{t+1}, a_{t+1}) - Q(s_t, a_t)] .$$

The convergence requirement for the Q-function computed in this way to approximate the optimal state-action function is that all pairs of state-action pairs will keep being updated. Under this assumption, the variants of the Q-learning algorithm with stochastic approximation conditions on the parameters have also been proved to converge with probability 1 to the optimal state-action function. The Q-learning algorithm is simply an iterative update on the Q-table with $|S|$ rows and $|A|$ columns, as shown in Figure 7.2.

```
1 function Q_Leaning(Episodes)
2    Initialize the Q(s,a), and Q(terminate-state,:)=0
3    for each episode in Episodes:
4        Initialize to start state S
5        repeat (for each step of episode):
6            choose an action a ∈ A using the policy from the Q table with ε -greedy
7            Take the action a and observe the reward R and the next state s'
8            Q(s,a) ← Q(s,a) + α[R + γ max Q(s',a') − Q(s,a)]
                                              a'
9            S ← s'
10       until S is a terminal state
```

Figure 7.2: Q-learning algorithm.

The ε-greedy choice of taking an action means that the agent randomly chooses an action to take with an ε probability, while taking the best possible action (the one that gives the best expected return given the current state s_t) with $1 - \varepsilon$ probability.

However, with the algorithm in Figure 7.2, the table size could become extremely large with a large state space $|S|$ and action space $|A|$. In the autonomous driving case, where the state space could be a combination of sensor input, vehicle status, and localization and map information, a state representation will be a large multi-dimensional vector where the value in each dimension will be continuous. Of course, one could argue that large continuous state spaces could always be discretized and we could maintain a big sparse Q-table. However, it may be extremely time costly for such a huge Q-table to converge. Further, learning may be highly inefficient and even impractical. When the state-action space grows excessively, deep learning can be used as a good approximation for the Q-function. DNNs are very good at *coming up* with good features for highly structured data with large size. Figure 7.3 shows two candidate structures for the Q-function neural network approximation. The network on the left takes states and actions as input and outputs the q value for any state-action pair. However, this structure is not efficient since for every state-action pair, a forward pass computation has to be performed. In practice, DNNs with a structure similar to the one on the right are being used. This network only takes states as input while and the network outputs candidate q values for every possible action a ∈ A, making it sufficient to do just one pass of forward computation.

Figure 7.4 shows a typical DNN, which is the one used in Google's DeepMind paper and referred to as Deep-Q-Learning (DQN). It is a typical neural network with convolutional layers and with last stages of fully connected layers. The most significant difference with typical neural networks in computer vision is the absence of pooling layers. Pooling layers makes the network location invariant. In our case, we do not want object location to be characteristic of its existence when doing object detection.

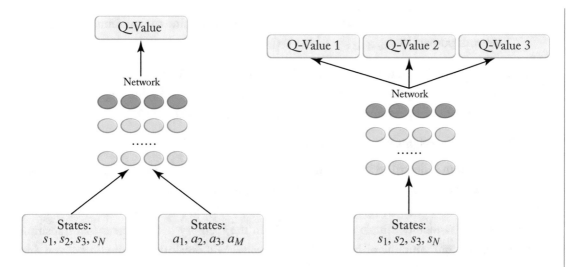

Figure 7.3: Deep-Q-Learning (DQN): Neural network structures.

Layer	Input	Filter Size	Stride	Num Filters	Activation	Output
conv1	84 × 84 × 4	8 × 8	4	32	ReLU	20 × 20 × 32
conv2	20 × 20 × 32	4 v 4	2	64	ReLU	9 × 9 × 64
conv3	9 × 9 × 64	3 × 3	1	64	ReLU	7 × 7 × 64
fc4	7 × 7 × 64			512	ReLU	512
fc5	512			18	Linear	18

Figure 7.4: Deep-Q-Learning: Typical neural network structure in DQN.

The network utilizes the structure on the right hand side of Figure 7.3, where the network outputs the q values of every possible action for any given state input. The training of the DNN could be viewed as a regression problem optimizing a squared error loss between predicted q values and target q values. The squared error loss can be expressed as:

$$L = \frac{1}{2}[r(s_t, a_t, s_{t+1}) + \gamma \max_{a_{t+1}} Q(s_{t+1}, a_{t+1}) - Q(s_t, a_t)]^2,$$

where $r(s_t, a_t, s_{t+1}) + \gamma \max_{a_{t+1}} Q(s_{t+1}, a_{t+1})$ is the target term and $Q(s_t, a_t)$ is the predicted term, given the transition (s_t, a_t, r, s_{t+1}). Correspondingly, the previous Q-function table-updating algorithm will be adapted to the procedure as follows.

1. Perform a forward pass for the current state s_t to get the predicted q values for all the actions.

2. Perform a forward pass for the next state s_{t+1} and determine the maximum over all network outputs $\max_{a_{t+1}} Q(s_{t+1}, a_{t+1})$.

3. Set the target q value for action a_t to be $r(s_t, a_t, s_{t+1}) + \gamma \max_{a_{t+1}} Q(s_{t+1}, a_{t+1})$, as calculated in the previous step. For every other action, set the q value target to be the same as originally returned in Step 1, making the error to be zero for those outputs.

4. Update the weights with back propagation.

The algorithm is described in detailed in Figure 7.5. In the algorithm, we used two methods to facilitate the learning process. The first is *experience replay*. The key idea is to store all the experienced state transitions (s_t, a_t, r, s_{t+1}) into a *replay memory*. When performing the training of the network, we then randomly sample mini-batches from the memory pool as the actual transition example. This random sampling of historical *replay memory* will break the distribution of training examples and introduce more variance in subsequent training examples. With experience replay, the training becomes supervised learning, which makes it easier to understand, debug, and test. If all the transition examples are from human replays, then training the DQN is closer to imitate human experiences.

```
1 function DQN_Leaning(Episodes)
2    Initialize the Q(s,a) with random weights
3    Initialize the replay memory D
4    for each episode in Episodes:
5        repeat :
6            choose an action a ∈ A using the policy from the Q table with ε -greedy
7            take the action a and observe the reward r and the next state s′
8            store experience (s, a, r, s′) in replay memory D
9            sample random transition (ss, aa, rr, ss′) from replay memory D
10           calculate target for each minibatch transition
11               if is terminal state:
12                   tt = rr
13               else:
14                   tt = rr + γ max Q(s′, a′) − Q(s, a)
                               a′
15               train deep neural network Q with Loss = ¹/₂ (tt − Q(ss, aa)) ²
16           s ← s′
17       until s is a terminal state
```

Figure 7.5: The Deep-Q-Learning algorithm.

The other method used to improve the performance of the Deep-Q-Learning algorithm is a balance of *exploration* and *exploitation* referred to as the ε-greedy action choice. The dilemma of exploration and exploitation lies in the fact that a good strategy picks actions which could bring more expected rewards. However, to find these effective actions, the policy has to first explore actions with unknown expected rewards. More specifically, *exploration* means that the agent tries new actions that have not been explored and are not guaranteed to yield good returns. In contrast, *exploitation* means that the agent selects the most effective action given in the current state according to the Q-table. Usually exploration happens at the beginning of the Q-learning process where the initial state-action pairs are randomly initialized. After the initial iterations, the agent has accumulated some knowledge about the expected returns for the state-action pairs. Thereafter, the agent will pick the "greedy" action with the highest Q-value given the state, which implies an emphasis on epxloiting the Q-table. Such balance is controlled by the ε-greedy exploration, with the parameter ε controlling the balance between exploration and exploitation. In DeepMind [4], the ε parameter is initially set to 1 to favor exploration at the beginning. It is then gradually decreased to shift toward exploitation, and finally settles down to a fixed exploration rate of 0.1.

7.2.2   ACTOR-CRITIC METHODS

One of the most significant breakthroughs in reinforcement learning algorithms is the *Asynchronous Advantage Actor-Critic* (A3C) algorithm presented by Google's DeepMind. It has been shown to be faster, simpler, and more robust than the traditional Deep-Q-Learning algorithms on standard reinforcement learning tasks.

The structure of an A3C network is shown in Figure 7.6. The most significant difference with DQN is that there are more than one learning agent in A3C as opposed to one for DQN. Each agent is represented as an individual neural network with its own parameters, interacting and learning within its own copy of the environment. These are the *worker agents*. In addition to these individual worker agents is the *global network agent*. The learning is asynchronous because each agent learns in its own environment. After each agent is informed of the loss, they update their policies in their own environments, and then update the global environment together. With the updated global network, each agent is reset to the updated global environment to kick off a new round of learning. This asynchronous learning process is more efficient than single-agent learning since every agent learns independently and the overall experience is more diverse. The detailed workflow for training an A3C algorithm is as follows.

1. Initialize a global network and reset each individual worker network to the global network.

2. Each individual worker agent interacts and learns within its own environment.

3. Each individual worker agent computes the loss function for its neural network.

4. Each individual worker agent updates its gradient from the computed loss.

5. Worker agents together update the global network with the appropriate gradients.

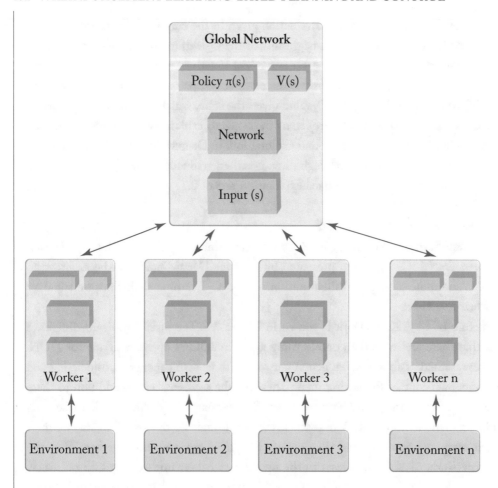

Figure 7.6: Asynchronous Advantage Actor-Critic Framework (based on [11]).

As shown in Figure 7.6, the neural networks in both the worker and global networks have two unique *branches*. These two branched networks are designed to separately estimate both the value function $V_\pi(s)$ and the policy $\pi(s)$. While the value function $V_\pi(s)$ represents the desirability to be in a state, the policy $\pi(s)$ estimates the desirability output for each action. These two separate value estimate and policy estimate networks have their own fully connected layers in their branch. In addition, the value network and the policy network share common network structures, which consist of convolutional layers to achieve location invariance, and an LSTM layer to handle temporal dependencies on top of the convolutional layers. The network branch for estimating value function $V_\pi(s)$ serves as the role of *critic*, while the other network branch for estimating policy $\pi(s)$ is the *actor*. Most importantly, the agent uses the value function estimate from the critic to update the policy function represented as the actor, which makes it a more effective way than general policy gradient methods. Hence, this specialized distinctive network branches used for separately estimating value and policy functions is referred to as an *Actor-Critic* method. Finally, another uniqueness of the A3C algorithm is its *Advantage* update. While the policy gradient method uses discounted expected returns, the A3C algorithm keeps pondering the question of "how much better I am than expected." The advantage function measures this quantity in the form of: $A = Q(s, a) - V(s)$, where the $Q(s, a)$ could be approximated by the discounted return R. The value loss function for each agent worker is represented as:

$$L_{value} = \sum (R - V(s))^2.$$

To achieve a balanced control between exploitation and exploration, the entropy of output action probabilities H_π is computed and incorporated into the loss function for policy updates. When the entropy is high, it means that output actions have small probabilities and the agent should be more conservative in exploitation and do more exploration. If the entropy is low, which indicates a confident policy, the agent then turns toward exploitation. The loss function for the policy is therefore:

$$L_{policy} = A(s) * \log(\pi(s)) + H_\pi * \beta.$$

Note that [11, 12] provide a detailed tensorflow-based A3C implementation.

7.3   LEARNING-BASED PLANNING AND CONTROL IN AUTONOMOUS DRIVING

With the description of the most popular reinforcement learning algorithms like Q-Learning and Actor-Critic methods, we now dive into how reinforcement learning algorithms have helped autonomous driving planning and control. Reinforcement learning has been applied to different levels of autonomous driving planning and control, including but not limited to levels of: *behavior decision*, *motion planning*, and *feedback control*. In another instance, [8] used DNN-based supervised learning in autonomous driving. The inputs are raw sensor data such as image pixels and the outputs are direct control signals such as steering, throttle, and brake. Such approaches are called *end-to-end* solutions. While the end-to-end ideas of *sensor data in*, *control signal out* sound very attractive, it is usually accompanied with complicated model structures, and the end results and/or intermediate results are very hard to explain if not totally impossible. In the following subsections, we introduce reinforcement learning-based approaches to various planning and control layers. The goal of this section is not to cover all the reinforcement learning-based planning and control approaches, but to iterate and discuss over the most typical approaches on different layers. Throughout this discussion, we will focus on the following aspects: problem extent, state space design, action space design, network structure design, as well as application constraints. In most of extent works, the output actions imply the extent to which reinforcement learning is performed. These actions usually remain within the level of behavior decision or direct control. Surprisingly, there are very few works where the output actions remain at the level of motion planning (i.e.. spatial-temporal trajectories). This shows that many of the reinforcement learning approaches dive deep into the actual spatial-temporal trajectory execution level at the bottom control layer.

7.3.1 REINFORCEMENT LEARNING ON BEHAVIORAL DECISION

The main goal for applying reinforcement learning in behavioral decisions is to tackle the highly diverse traffic scenarios where simply literally following traffic rules could not be very helpful. To tackle the "long-tail" cases in behavioral decisions, human driving experiences can serve as excellent examples in teaching a reinforcement learning-based system to make more human-like decisions. This could be a very good complement to a rule-based behavioral decision approach which remains the mainstream industrial approach. In [3], a reinforcement learning-based approach is applied on the behavior decision level, where the action space (referred to as *Desires* in the paper) is defined as:

$$D = [0, v_{max}] \times L \times \{g, t, o\}^n,$$

where v_{max} is the desired target speed of the autonomous vehicle, L is a discrete set of lateral lane positions, and g, t, o, respectively represent *give way to* (yield), *take* (overtake), and *keep an offset distance* (nudge/attention) for other obstacle vehicles. The action space (*Desires*) is the Cartesian product of these three dimensions. The state space for the approach in [3] contains the "environment model" around the vehicle generated from interpreting sensory information as well as any additional useful information such as kinematics of moving objects from previous frames.

One of the key contributions for this reinforcement learning-based decision strategy is that the stochastic gradient of the policy does not necessarily requires to obey the Markov property. Therefore, methods which reduce the variance of the gradient estimator would not require Markov assumptions either. While their implementation is proprietary and the results could not be reproduced, the authors mentioned that they initialized the reinforcement learning agent with imitation and updated it using an iterative Policy Gradient approach.

7.3.2 REINFORCEMENT LEARNING ON PLANNING AND CONTROL

The key challenge in reinforcement learning-based planning and control is how to design the state spaces. To compute the motion planning or feedback control level actions, it is necessary to include both autonomous vehicle information and the surrounding environment. If we do not take the raw sensory data as input, the state spaces will have to somehow incorporate structured information about the host autonomous vehicle and its environment. Thus, the state space will be a large multi-dimensional continuous space. To tackle the challenge of a continuous state space, cell-mapping techniques can be combined with reinforcement learning to solve the control problem of Car-Like-Vehicles (CLV) in [2]. The proposed method in [2] discretizes the state space with cell-mapping techniques, where the adjoining property is placed as a constraint for state transitions. The state space (before cell-mapping) and action space are shown in Figure 7.7.

State Symbol	State Variable	Range
$X_1 = v$	Velocity	$-1.5 \le X_1 \le 1.5$ "m/s"
$X_2 = x$	X Cartesian Coordinate	$-0.9 \le X_2 \le 0.9$ "m"
$X_2 = y$	Y Cartesian Coordinate	$-1.3 \le X_3 \le 1.3$ "m"
$X_2 = \theta$	Orientation	$-\pi \le X_4 \le \pi$ "rad"

Action Symbol	Action Values		
Voltage in traction motor	-18 V	0 V	18 V
Steering angle	$-23°$	$0°$	$23°$

Algorithm 1 CACM-RL

```
1:    Initialise Q-Table(s,a) y Model_Table
2:    x ← current state
3:    s ← cell(x)
4:    IF      s ∈ drain or s ∈ goal or s ∈ safety_area
5:    THEN  F_reactive(x)
6:    ELSE  IF      D-k-adjoining(x,x')
7:            THEN  Q-Table(s,a)← s',r
8:                  Model_Table ← IT(x,x')
9:            a ← policy(s)
10:           Execute action a on the vehicle
11:           Observe the new state x' and r
12:   UNTIL the end of the learning stage
13:   FOR all (s,a), repeat N times
14:       x̄' ← Model_Table, DT(x,x')
15:       s̄' ← cell(x̄')
16:       Q-Table(s,a) ← s̄',r
```

Figure 7.7: State space before cell-mapping, action space, and the Control-Adjoining-Cell-Mapping algorithm in [2], used with permission.

This approach does not use any neural network structure for the reinforcement learning part since the adjoining property significantly reduces the actual state spaces. Instead, a Q-learning fashioned table-updating algorithm is used, as shown in Figure 7.7. One important characteristic of this reinforcement learning-based approach is that it incorporates two tables of state-action pairs, the Q-table and the model-table. While the Q-table has the same meaning as traditional reinforcement learning, the model-table maintains an average of local transitions that satisfy the D-k adjoining property such that a good approximation to the optimal control policy could be represented. In addition, the CACM-RL algorithm in Figure 7.7 only performs exploration since the strict adjoining cell-to-cell mapping makes it unnecessary to do exploitation. One thing to mention is that obstacle information is only considered into the *F_reactive()* and the *safety_area* judgment function, as shown in Figure 7.7, and is not considered into any state variables. This choice greatly reduces the complexity of the state space, but it makes the algorithm not very robust against dynamic obstacles, leaving an opening for future work applying reinforcement learning in the general scenario of static/dynamic obstacle avoidance.

Special Scenarios

The two typical examples described above utilize reinforcement learning in order to tackle behavioral decision or feedback control problems in a general sense, since their output action space is not customized to any specific scenarios. In [1], recurrent neural network- (RNN) based approaches are tailored to handle the control problems in two special cases: adaptive cruise control (ACC) and merging into a traffic circle. While this approach customized for these two special cases is not generally adoptable, it does evoke an interesting idea of: "predict the predictable near future and learn the unpredictable environment." In both cases, the problem is decomposed into two phases. The first one is a supervised learning problem where a differentiable function $\hat{N}(s_t, a_t) \approx s_{t+1}$ mapping the current state s_t and action a_t to the next state s_{t+1} is learned. The learned function is a predictor for the short-term near future. Then, a policy function mapping from state space S to action space A is defined as a parametric function $\pi_\theta: S \rightarrow A$, where π_θ is an RNN. The next state s_{t+1} is defined as: $s_{t+1} = \hat{N}(s_t, a_t) + v_t$, where $v_t \in \mathbb{R}^d$ is the unpredictable environment. The second phase of the problem is to learn the parameter of π_θ by back-propagation. Given that v_t expresses the unpredictable aspects of the environment, the proposed RNN will learn a robust behavior invariant to the adversarial environment.

Some Thoughts on Unsolved Problems and Challenges

With these existing works on reinforcement learning-based planning and control, let us discuss several key challenges that must be tackled if we are to successfully apply a reinforcement learning-based behavioral decision or motion planning.

The first challenge is the design of the state space. State design is especially challenging if we want to incorporate environmental information like surrounding dynamic obstacles, as well as structured map information. Given that the number of nearby obstacles is not fixed, there has to be some rules to organize them if we want to include them into the state vector space. Moreover, information about an obstacle itself is not sufficient, and we would also need information about the obstacle's association with road structure. Considering all this information together is necessary for a general-purpose decision or planning. One way to include such information might be to divide the road-map into grids on the SL-coordinate system, which is similar to what the motion planning optimization does while searching for an optimal path. It is also an indicator that the way optimization-based motion planning handles the input data could be leveraged as a heuristic to design state spaces in a reinforcement learning-based approach.

How to design the reward function is another challenge. If the action output space concerns behavioral decisions that are mostly categorical, it might be easier for reward function design. However, for actual control-related action output and state space, how to measure a reward by a transition from state s_t to state s_{t+1} taking action a_t has to be carefully calibrated. Such a reward system needs to consider factors such as: reaching the destination, avoiding obstacles, and ride comfort. In fact, it might be easier to design such reward functions directly at the feedback control signal level rather than at the motion planning spatial-temporal trajectory level, since more past work has had a direct bearing on reinforcement learning on control signal outputs rather than on the motion planning trajectory. However, we do believe that performing learning-based on the motion planning trajectory level is appropriate because the module division in our planning and control framework separates the control layer as simply a feedback closed-loop system to execute the motion planning spatial-temporal output trajectories. Also, the problem of executing a spatial-temporal trajectory given a certain control system in a closed-loop could be purely modeled as an optimization problem, either through a linear or nonlinear system. Performing reinforcement learning directly at the control level might ease the problem of reward function design. However, the learning has to tackle the motion execution problem that should be mostly addressed by optimization in the control area. This implies that the learned model will have to unnecessarily model the trajectory execution rather than focus on generating an optimal trajectory itself. More efforts with reinforcement learning on autonomous driving motion planning will help.

7.4   CONCLUSIONS

We believe that solving the autonomous driving planning and control problems via reinforcement learning is an important trend and a critical complement to optimization-based solutions. With large amounts of accumulated driving data, reinforcement learning-based planning and control will help solve cases that optimization solutions alone cannot fully and reliably address. Current works on reinforcement learning techniques have been applied to different extents in autonomous driving planning and control, but much of it is still at an early stage.

7.5   REFERENCES

[1]   Shai, S-S., Ben-Zrihem, N., Cohen, A., and Shashua, A. 2016. *Long-term Planning by Short-term Prediction*. arXiv preprint arXiv:1602.01580. 143, 145, 156

[2]   Gómez, M., González, R. V., Martínez-Marín T., Meziat D., and Sánchez S. 2012. Optimal motion planning by reinforcement learning in autonomous mobile vehicles. *Robotica*, 30(2), pp. 159–70. DOI: 10.1017/S0263574711000452. 143, 145, 155

[3]   Shalev-Shwartz, S., Shammah, S., and Shashua, A. 2016. *Safe, Multi-agent, Reinforcement Learning for Autonomous Driving*. arXiv preprint arXiv:1610.03295. 143, 144, 154

[4]   Mnih, V., Kavukcuoglu, K., Silver, D., Graves, A., Antonoglou, I., Wierstra, D., and Riedmiller M. 2013. Playing Atari with deep reinforcement learning. arXiv preprint arXiv:1312.5602. 151

[5]   Katrakazas, C., Quddus, M., Chen, W-H., and Deka, L. 2015. Real-time motion plan-
        ning methods for autonomous on-road driving: State-of-the-art and future research
        directions. *Elsevier Transporation Research Park C: Emerging Technologies*, 60, pp. 416–442
        DOI: 10.1016/j.trc.2015.09.011. 143

[6]   Paden, B., Cap, M., Yong, S. Z., Yershow, D., and FrazzoloE. 2016. A survey of motion
        planning and control techniques for self-driving urban vehicles. *IEEE Transactions on
        Intelligent Vehicles*, 1(1), pp. 33–55. DOI: 10.1109/TIV.2016.2578706. 143

[7]   Sutton, R. S. and Barto, A. G. 1998. *Reinforcement Learning: An Introduction*. Cambridge:
        MIT Press. 147

[8]   Bojarski, M., Del Testa, D., Dworakowski, D., Firner, B., Flepp, B., Goyal, P., Jackel, L.
        D., Monfort, M., Muller, U., Zhang, J., and Zhang, X. 2016. *End to End Learning for
        Self-driving Cars*. arXiv preprint arXiv:1604.07316. 143, 153

[9]   Geng, X., Liang, H., Yu, B., Zhao, P., He, L., and Huang, R. 2017. A scenario-adaptive
        driving behavior prediction approach to urban autonomous driving. *Applied Sciences*, 7(4),
        p. 426. DOI: 10.3390/app7040426. 143, 144

[10]   *SAE Levels of Driving Automation*. https://www.sae.org/misc/pdfs/automated_driving.
        pdf. 144

[11]   Mnih, V., Badia, A. P., Mirza, M., Graves, A., Lillicrap, T., Harley, T., Silver, D., and
        Kavukcuoglu, K. 2016. Asynchronous methods for deep reinforcement learning. In *In-
        ternational Conference on Machine Learning* (pp. 1928–1937). 152

[12]   https://github.com/awjuliani/DeepRL-Agents/blob/master/A3C-Doom.ipynb.

CHAPTER 8

Client Systems for Autonomous Driving

Abstract

This chapter focuses on the design of autonomous driving client systems, including the operating system and the computing platform. An autonomous system is a very complex software and hardware system. In order to coordinate the interactions between different components, an operating system is required, and the operating system we discuss in this chapter is based on the Robot Operating System (ROS). Next we discuss the computing platform, which is the brain of this complex autonomous driving system. We will analyze the computational requirements of autonomous driving tasks, the advantages and disadvantages of each autonomous driving computing solutions, like CPU, GPU, FPGA, DSP, and ASIC.

8.1 AUTONOMOUS DRIVING: A COMPLEX SYSTEM

Autonomous driving is a highly complex system that consists of many different tasks. As shown in Figure 8.1, in order to achieve autonomous operation in urban situations with unpredictable traffic, several real-time systems must interoperate, including sensor processing, perception, localization, planning and control.

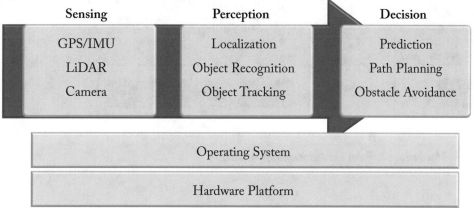

Figure 8.1: Hardware platform for autonomous driving.

Autonomous driving is not one technology but an integration of many technologies, and the integration is done at the level of the client system, which consists of the operating system and the hardware platform. Figure 8.2 below shows a greatly simplified version of the hardware platform, it follows the sensing, perception, and action computing paradigm introduced in Chapter 1. First the sensors collect data from the environment and feed these data to the computing platform for perception and action computation, then the action plans are sent to the control platform for execution. Having the hardware itself is not sufficient; on top of the hardware, we need an operating system to coordinate all the communications between these components, as well as to coordinate the resource allocation for different real-time tasks. For instance, the camera needs to deliver 60 frames per second, implying that the processing time for each frame should be less than 16 ms. When the amount of data increases, the allocation of system resources becomes a problem: for example, when a burst of LiDAR point cloud data gets into the system, it could severely contend for CPU resources, thus leading to dropped frames on the camera side. Therefore, we need a mechanism to restrict the amount of resources used by each component, which is one of the mains tasks of the operating system.

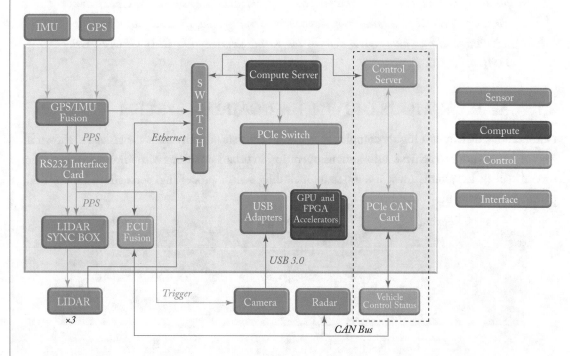

Figure 8.2: Hardware platform for autonomous driving.

8.2   OPERATING SYSTEM FOR AUTONOMOUS DRIVING

Autonomous systems integration includes multiple software modules: sensing, localization, object recognition, object tracking, traffic prediction, path planning, obstacle avoidance, navigation, etc. Each of these components need to meet some real-time requirements in order for the autonomous vehicle to work. Therefore, we need an operating system to manage all these components. The two main functions provided by the operating system include communication, and resource allocation. The Robot Operating System (ROS) is a set of software libraries and tools that provides these capabilities [1], and to our knowledge, many production autonomous driving operating systems either use ROS directly or apply the ROS design philosophy. Therefore, we start our discussion with ROS.

8.2.1   ROS OVERVIEW

ROS originated in the Willow Garage PR2 project. The main components are divided into 3 types: ROS Master, ROS Node, and ROS Service. The main function of ROS master is to provide name service. It stores the operating parameters that are required at startup, the name of the connection between the upstream node and the downstream node, and the name of the existing ROS services. The ROS nodes process the received messages and releases the new message to the downstream nodes. The ROS service is a special ROS node, which is equivalent to a service node, accept the request and return the results of the request. The second generation of ROS, ROS 2 is optimized for industrial applications, it uses the DDS middleware for reliable communication and and it uses shared memory to improve communication efficiency (Figure 8.3).

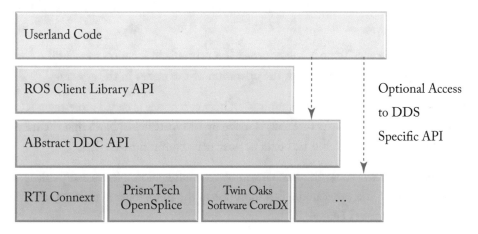

Figure 8.3: ROS 2.0 communication with DDS.

However, unlike ROS 1.0, which has been heavily tested for years, ROS 2.0 is still under development. Thus, most autonomous driving operating systems still rely on ROS 1.0 instead of ROS 2.0 for the following reasons.

1. Stability and security are critical in an autonomous environment. We need to ensure the stability and security of the system by using a proven stable system, but ROS 2.0 is still unproven.

2. The cost of DDS itself. We tested the performance cost of using DDS middleware and found out that the throughput of DDS is even worse than that of ROS 1.0. The main reason being that the overheads of using DDS is quite high.

The Basics of ROS

The most important concepts in ROS include node, node manager, parameter server, message, theme, service, and task.

1. **Node:** A node is a process used to perform a task. For instance, the motor control node is used to read the motor information and control the motor rotation. The path-planning node is used to realize the motion planning of mobile platforms.

2. **Node manager/Master:** As the name implies, the purpose of the node manager is to manage other nodes. Each node needs to register its information with the node manager, so that the node manager can coordinate the communications between the nodes.

3. **Parameter server:** The parameter server is a centralized location to store the configuration parameters required for the operation of the nodes in the system.

4. **Message:** The content of communication between nodes is called a message. A message is a simple data structure that is made up of typed fields. Note that the message can encapsulate structured text data or unstructured multimedia data.

5. **Topic:** A publish-subscribe mechanism of communication. Nodes can publish message to a topic and other nodes can subscribe to the same topic in order to receive the published messages (Figure 8.4).

6. **Service:** A one-to-one communication mechanism. A node can request the service provided by a service node, and as a result a communication channel is established between these two nodes (Figure 8.4).

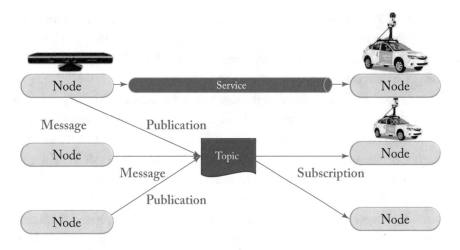

Figure 8.4: **ROS** communication mechanisms.

8.2.2   SYSTEM RELIABILITY

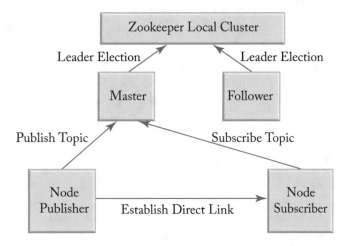

Figure 8.5: **ROS** communication mechanisms.

As mentioned in the beginning of this section, system reliability is the most important requirement for the autonomous driving operating system. Imagine, while an autonomous vehicle is moving on the road, suddenly the ROS master node crashes, leading to a system shutdown. This scenario is likely to happen in the original ROS design, as there is only one master in the whole system, but it is not acceptable in autonomous vehicle applications. Thus, the first task is to decentralize the master node to achieve robustness and reliability. As shown in Figure 8.5, one way to solve this problem is to utilize ZooKeeper [2], with which multiple master nodes are maintained, one serving as the active master node, and others serving as the backup master nodes. In this case, when the active master node crashes, one of the backup master nodes will be elected as the new active node, thus preventing a system crash.

The ZooKeeper mechanism handles the case of master node crashes, however, other nodes, such as the planning node, may crash too. To handle this scenario, we implemented a monitor node in the system to keep track of the status of all nodes. In this system, each node in the system sends a periodic heartbeat message to the ZooKeeper, and if a heartbeat is not detected for a period of time, then we can assume that the node is lost. Then the ZooKeeper notifies the monitor node to restart the lost node. At times when we restart a node, the node is stateless, for instance, a node that processes incoming images does not have to keep state. At other times, some nodes need to keep track of the state, such as the localization node needs to know its current position, then when we restart we need to restart from the last checked-in state. Therefore, when a heartbeat is sent to the ZooKeeper, sometimes it contains the state information of the sending node, and we can use this last known state information to restart the lost node.

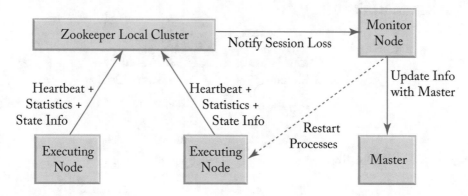

Figure 8.6: Monitor node.

8.2.3   PERFORMANCE IMPROVEMENT

In the original ROS design, when communication is frequent, it introduces high performance overhead. First, the communication between ROS nodes in the same machine applies a loop-back mechanism through the whole network stack. That is to say, every packet needs to be processed through a multi-layer software stack, causing unnecessary delay and memory consumption. In order to solve this problem, we applied shared memory and memory-mapping techniques to force the data to be communicated through the memory system and then sending the address only. Applying this technique, for local communication, we only need send one packet and therefore bound the communicate latency to within 20 microseconds, significantly reducing CPU usage.

Second, when a broadcast is performed in the original ROS, the underlying implementation actually uses multiple point-to-point transmissions. For instance, if one wanted to send data to five nodes, the same data would be copied five times, one for each node. This would cause a significant waste of computing and memory resources. In addition, it would impose severe demands on the throughput of the communication system. In order to address this problem, we implemented a multicast mechanism such that if a sending node sends a message to multiple receiving nodes, only one copy of the message is needed.

Third, after studying the communication stack of ROS, we found that the communication latency is mostly caused by data serialization and deserialization. Serialization is the process that converts the state information of an object into a form that can be stored or transmitted. During serialization, the object writes its current state to a temporary or persistent store. After that, wo can recreate the object by reading the state of the object from the storage area. To solve this problem, we implemented a lightweight serialization/deserialization mechanism that reduces the serialization latency by 50%.

8.2.4 RESOURCE MANAGEMENT AND SECURITY

We can now imagine one simple attack scenario, where one of the ROS nodes is hijacked, and then it keeps allocating memory until the system hits the Out-Of-Memory (OOM) error and starts killing processes to free up memory. This case can actually happen in the original ROS design as there is no security mechanism in place to prevent this. To approach this problem, we encapsulated each ROS node with Linux Container (LXC) [3]. In short, LXC provides lightweight virtualization, so as to isolate the process and resources. Similar to C++ NameSpace, LXC effectively divides the resources into isolated groups to constrain the resources that can be used by each node, so that guarantees that each node has enough computing and memory resources to meet the real-time requirements. Moreover, as a virtualization technique, LXC is lightweight and only brings about 5% of CPU overheads. In addition to resource constraints, LXC also provides sandbox support, allowing the system to limit the permissions of the ROS node process. In order to avoid potentially dangerous ROS node processes that may destroy other ROS node processes, Sandbox technology can restrict access to disk, memory, and network resources. Hence, using LXC, we provide not only security to the ROS nodes, but also a means to allocate and manage system resources.

8.3 COMPUTING PLATFORM

On a production autonomous vehicle, each second, the sensor can generate as much as 2 GB of raw sensor data, and then the enormous amount of data is fed into the computing platform for perception and action plan computation. Therefore, the design of the computing platform directly affects the real-time performance as well as robustness of the autonomous driving system. The key issues include cost, power consumption, heat dissipation, etc.

8.3.1   COMPUTING PLATFORM IMPLEMENTATION

To start, we first take a look at an existing autonomous driving computing solution, provided by a leading autonomous driving development company [4]. The first generation of this computing platform consists of two compute boxes, each equipped with an Intel Xeon E5 processor and four to eight Nvidia K80 GPU accelerators, connected with a PCI-E bus. At its peak performance, the CPU (which consists of 12 cores), is capable of delivering 400 GOPS/s, consumes 400 W of power. Each GPU is capable of 8 TOPS/s, while consuming 300 W of power. Combining everything together, the whole system is able to deliver 64.5 TOPS/s at about 3,000 W. The compute box is connected to 12 high-definition cameras around the vehicle, for object detection and object tracking tasks. A LiDAR unit is mounted on top of the vehicle for vehicle localization as well as some obstacle avoidance functions. A second compute box performs exactly the same tasks and is used for reliability: in case the first box fails, the second box can immediately take over. In the worst case, when both boxes run at their peak, this would mean over 5,000 W of power consumption that would consequently generate enormous amount of heat. Also, each box costs $20,000–$30,000, making the whole solution unaffordable to average consumers.

8.3.2   EXISTING COMPUTING SOLUTIONS

In this subsection, we will present the existing computing solutions provided by chip designers and manufacturers for autonomous driving computing.

GPU-based computing solution

The Nvidia PX platform is the current leading GPU-based solution for autonomous driving. Each PX 2 consists of two Tegra SoCs and two Pascal graphics processors. Each GPU has its own dedicated memory, as well as specialized instructions for Deep Neural Network acceleration. To deliver high throughput, each Tegra connects directly to the Pascal GPU using a PCI-E Gen 2 × 4 bus (total bandwidth: 4.0 GB/s). In addition, the dual CPU-GPU cluster is connected over Gigabit Ethernet, delivering 70 GB/s per second. With optimized I/O architecture and DNN acceleration, each PX2 is able to perform 24 trillion deep-learning calculations every second. This means that, when running AlexNet deep learning workloads, it is capable of processing 2,800 images/s.

DSP-based solution

Texas Instruments' TDA provides a DSP-based solution for autonomous driving. A TDA2x SoC consists of two floating-point C66x DSP cores and four fully programmable Vision Accelerators, which are designed for vision processing functions. The Vision Accelerators provide eight-fold acceleration on vision tasks compared to an ARM Cortex-15 CPU, while consuming less power. Similarly, CEVA XM4 is another DSP-based autonomous driving computing solution. It is designed for computer vision tasks on video streams. The main benefit for using CEVA-XM4 is energy-efficiency, which requires less than 30 mW for a 1080p video at 30 frames per second.

FPGA-based solution

Altera's Cyclone V SoC is one FPGA-based autonomous driving solution that has been used in Audi products. Altera's FPGAs are optimized for sensor fusion, combining data from multiple sensors in the vehicle for highly reliable object detection. Similarly, Zynq UltraScale MPSoC is also designed for autonomous driving tasks. When running Convolution Neural Network tasks, it achieves 14 images/sec/Watt, which outperforms the Tesla K40 GPU (4 images/s/Watt). Also, for object tracking tasks, it reaches 60 fps in a live 1080p video stream.

ASIC-based solution

MobilEye EyeQ5 is a leading ASIC-based solution for autonomous driving. EyeQ5 features heterogeneous, fully programmable accelerators, where each of the four accelerator types in the chip are optimized for their own family of algorithms, including computer-vision, signal-processing, and machine-learning tasks. This diversity of accelerator architectures enables applications to save both computational time and energy by using the most suitable core for every task. To enable system expansion with multiple EyeQ5 devices, EyeQ5 implements two PCI-E ports for inter-processor communication.

8.3.3 COMPUTER ARCHITECTURE DESIGN EXPLORATION

We attempt to develop some initial understandings of the following questions: (1) What computing units are best suited for what kind of workloads? (2) If we went to the extreme, would a mobile processor be enough to perform the tasks in autonomous driving? (3) How to design an efficient computing platform for autonomous driving?

Matching Workloads to Computing Units

We try to understand which computing units fit best for convolution and feature extraction workloads, which are the most computation-intensive workloads in autonomous driving scenarios. We conducted experiments on an off-the-shelf ARM mobile SoC consisting of a four-core CPU, a GPU, as well as a DSP, the detailed specifications can be found in [5]. To study the performance and energy consumption of this heterogeneous platform, we implemented and optimized feature extraction and convolution tasks on CPU, GPU, and DSP, and measured chip-level energy consumption.

First, we implemented a convolution layer, which is commonly used, and is the most computation-intensive stage, in object recognition and object tracking tasks. The left side of Figure 8.7 summarizes the performance and energy consumption results: when running on the CPU, each convolution takes about 8 ms to complete, consuming 20 mJ; when running on the DSP, each convolution takes 5 ms to complete, consuming 7.5 mJ; when running on a GPU, each convolution takes only 2 ms to complete, consuming only 4.5 mJ. These results confirm that GPU is the most efficient computing unit for convolution tasks, both in performance and in energy consumption.

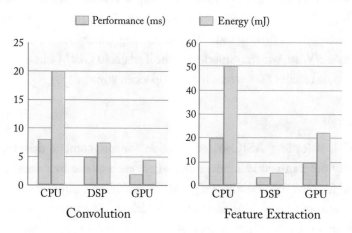

Figure 8.7: Convolution and feature extraction performance and energy.

Next, we implemented feature extraction, which generates feature points for the localization stage, and this is the most computation expensive task in the localization pipeline. The right side of Figure 8.7 summarizes the performance and energy consumption results: when running on a CPU, each feature extraction task takes about 20 ms to complete, consuming 50 mJ; when running on GPU, each convolution takes 10 ms to complete, consuming 22.5 mJ; when running on a DSP, each convolution takes only 4 ms to complete, consuming only 6 mJ. These results confirm that DSP is the most efficient computing unit for feature processing tasks, both in performance and in energy consumption. Note that we did not implement other tasks in autonomous driving, such as localization, planning, obstacle avoidance, etc., on GPUs and DSPs as these tasks are control-heavy and would not efficiently execute on GPUs and DSPs.

Autonomous Driving on Mobile Processor

We seek to explore the edges of the envelope and understand how well an autonomous driving system could perform on the aforementioned ARM mobile SoC. Figure 8.8 shows the vision-based autonomous driving system we implemented on this mobile SoC. We implemented on this mobile SoC, we utilize the DSP for sensor data processing tasks, such as feature extraction and optical flow; we use GPU for deep learning tasks, such as object recognition; we use two CPU threads for localization tasks to localize the vehicle at real-time; we use one CPU thread for real-time path planning; and we use one CPU thread for obstacle avoidance. Note that multiple CPU threads can run on the same CPU core if a CPU core is not fully utilized.

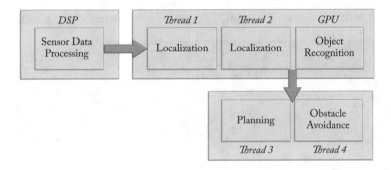

Figure 8.8: Autonomous navigation system on mobile SoC.

The performance was quite reasonable when we ran this system on the ARM Mobile SoC. The localization pipeline is capable of processing 25 images per second, almost keeping up with image generation at 30 images per second. The deep learning pipeline is capable of performing 2 to 3 object recognition tasks per second. The planning and control pipeline is designed to plan a path within 6 ms. When running this full system, the SoC consumes 11 W on average. With this system, we were able to drive the vehicle at around 5 mph without any loss of localization, quite a remarkable feat, considering that this ran on a mobile SoC. With more computing resources, the system should be capable of processing more data and allowing the vehicle to move at a higher speed, eventually satisfying the need of a production-level autonomous driving system.

Design of Computing Platform

The reason why we could deliver this performance on an ARM mobile SoC is that we fully utilized the heterogeneous computing resources of the system and used the best suited computing unit for each task so as to achieve best possible performance and energy efficiency. However, there is a downside as well: we could not fit all the tasks into such a system, for example, object tracking, change lane prediction, cross-road traffic prediction, etc. In addition, we aim for the autonomous driving system to have the capability to upload raw sensor data and processed data to the cloud, however, the amount of data is so large that it will take all the available network bandwidth.

The aforementioned functions, object tracking, change lane prediction, cross-road traffic prediction, data uploading etc. are not needed all the time. For example, the object tracking task is triggered by the object recognition task and the traffic prediction task is triggered by the object tracking task. The data uploading task is not needed all the time either since uploading data in batches usually improves throughput and reduces bandwidth usage. If we designed an ASIC chip for each of these tasks, it would be a waste of chip area, but an FPGA would be a perfect fit for these tasks. We could have one FPGA chip in the system and have these tasks time-share the FPGA. It has been demonstrated that using Partial-Reconfiguration techniques [6], an FPGA soft core could be changed within less than a few milliseconds, making time-sharing possible in real time.

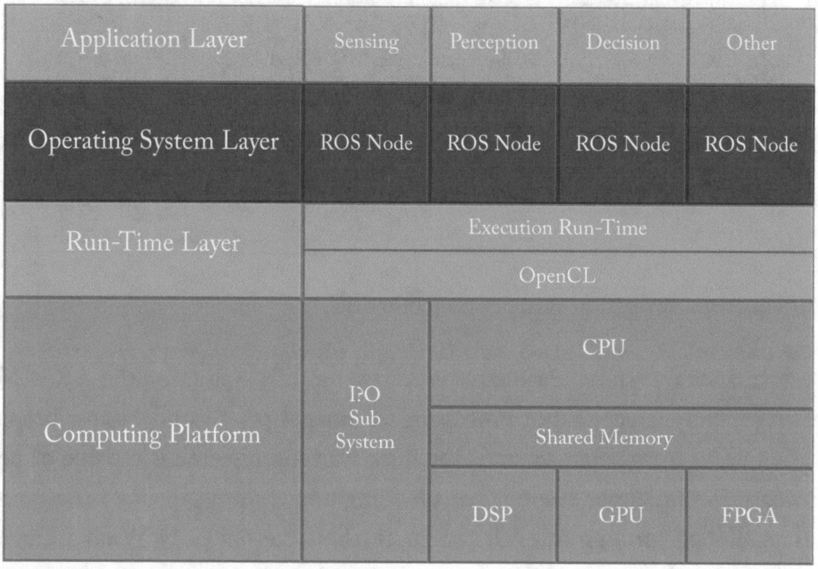

Application Layer	Sensing	Perception	Decision	Other
Operating System Layer	ROS Node	ROS Node	ROS Node	ROS Node
Run-Time Layer	Execution Run-Time			
	OpenCL			
Computing Platform	I?O Sub System	CPU		
		Shared Memory		
		DSP	GPU	FPGA

Figure 8.9: **Computing stack for autonomous driving.**

In Figure 8.9, we show our computing stack for autonomous driving. At the computing platform layer, we have a SoC architecture that consists of a I/O sub-system that interacts with the front-end sensors; a DSP to pre-process the image stream to extract features; a GPU to perform object recognition and some other deep learning tasks; a multi-core CPU for planning, control, and interaction tasks; an FPGA that can be dynamically reconfigured and time-shared for data compression and uploading, object tracking, and traffic prediction, etc. These computing and I/O components communicate through the shared memory. On top of the computing platform layer, we have a run-time layer to map different workloads to the heterogeneous computing units through OpenCL [7], and to schedule different tasks at runtime with a run-time execution engine. On top of the Run-Time Layer, we have an Operating Systems Layer utilizing Robot Operating System (ROS) design, which is a distributed system consisting of multiple ROS nodes, each encapsulates a task in autonomous driving.

8.4   REFERENCES

[1]   Quigley, M., Gerkey, B., Conley, K., Faust, J., Foote, T., Leibs, J., Berger, E., Wheeler, R., and Ng, A. 2009. ROS: An open-source robot operating system, *Proceedings of the Open-Source Software Workshop International Conference on Robotics and Automation.* 161

[2]   Hunt, P., Konar, M., Junqueira, F. P., and Reed, B. 2010. ZooKeeper: Wait-free coordination for internet-scale systems. In *Proceedings of the Usenix Annual Technical Conference.* 164

[3]   Helsley, M. 2009. LXC: Linux container tools. *IBM DevloperWorks Technical Library*, p.11. 165

[4]   Liu, S., Tang, J., Zhang, Z., and Gaudiot, J. L. 2017. Computer architectures for autonomous driving. *Computer*, 50(8), pp.18–25. DOI: 10.1109/MC.2017.3001256. 166

[5]   *Qualcomm Snapdragon 820 Processor*, https://www.qualcomm.com/products/snapdragon/processors/820. 167

[6]   Liu, S., Pittman, R. N., Forin, A., and Gaudiot, J.L. 2013. Achieving energy efficiency through runtime partial reconfiguration on reconfigurable systems. *ACM Transactions on Embedded Computing Systems (TECS)*, 12(3), p. 72. DOI: 10.1145/2442116.2442122. 170

[7]   Stone, J. E., Gohara, D., and Shi, G. 2010. OpenCL: A parallel programming standard for heterogeneous computing systems. *Computing in Science & Engineering*, 12(3), pp. 66–73. DOI: 10.1109/MCSE.2010.69. 170

CHAPTER 9

Cloud Platform for Autonomous Driving

Abstract

Autonomous driving clouds provide essential services to support autonomous vehicles. Today these services include but not limited to distributed simulation tests for new algorithm deployment, offline deep learning model training, HD map generation, etc. These services require infrastructure support including distributed computing, distributed storage, as well as heterogeneous computing. In this chapter, we present the details of our implementation of a unified autonomous driving cloud infrastructure, and how we support these services on top of this infrastructure.

9.1 INTRODUCTION

Autonomous vehicles are mobile systems, and autonomous driving clouds provide some basic infrastructure supports including distributed computing, distributed storage, and heterogeneous computing [1]. On top of this infrastructure, we can implement essential services to support autonomous vehicles. For instance, as autonomous vehicles travel around a city, each second over 2GB of raw sensor data can be generated. It thus behooves us to create an efficient cloud infrastructure to store, process, and make sense of the enormous amount of raw data. With the cloud infrastructure introduced in this chapter, we can efficiently utilize the raw data to perform distributed simulation tests for new algorithm deployment, to perform offline deep learning model training, as well as to continuously generate HD maps.

9.2 INFRASTRUCTURE

The key cloud computing applications for autonomous driving include but are not limited to simulation tests for new algorithm deployment, HD map generation, offline deep learning model training, etc. These applications all require infrastructural support, such as distributed computing and storage. One way to do this is to tailor an infrastructure to each application, at the cost of several practical problems.

- **Lack of dynamic resource sharing:** If we tailored each infrastructure to one application, then we could not use them interchangeably even when one is idle and the other is fully loaded.

- **Performance degradation:** Data is sometimes shared across applications. For instance, a newly generated map can be used in the driving simulation workloads. Without a unified infrastructure, we often need to copy data from one distributed storage element to another, leading to high performance overhead.

- **Management overheads:** It may take a team of engineers to maintain each specialized infrastructure. By unifying the infrastructure, we would greatly reduce the management overhead.

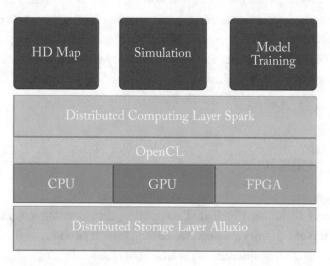

Figure 9.1: Cloud platform for autonomous driving.

As shown in Figure 9.1, to address these problems, we developed a unified infrastructure to provide distributed computing and distributed storage capabilities. To further improve performance, we built a heterogeneous computing layer to accelerate different kernels on GPUs or FPGAs, which either provide better performance or energy efficiency. We use Spark for distributed computing [2], OpenCL for heterogeneous computing acceleration [4], and Alluxio for in-memory storage [3]. By combining the advantages of these three infrastructure components, we can deliver a reliable, low-latency, and high-throughput autonomous driving cloud.

9.2.1   DISTRIBUTED COMPUTING FRAMEWORK

When we started building the distributed computing framework for autonomous driving, we had two options, the Hadoop MapReduce engine [11], which has a proven track record, or Apache Spark [2], an in-memory distributed computing framework that provides low latency and high throughput.

Specifically, Apache Spark provides programmers with an application programming interface centered on a data structure called the resilient distributed dataset (RDD), a read-only multiset of data items distributed over a cluster of machines maintained in a fault-tolerant way. It was developed in response to limitations in the MapReduce cluster computing paradigm, which forces a particular linear dataflow structure on distributed programs: MapReduce programs read input data from disk, map a function across the data, reduce the results of the map, and store reduction results on disk. In contrast, Spark's RDDs function as a working set for distributed programs that offer a restricted form of distributed shared memory. By using in-memory RDD, Spark can reduce the latency of iterative computation by several orders of magnitude.

Before switching to Spark from MapReduce, we focused on the reliability of the Spark cluster to determine whether it can deliver the needed performance improvement. First, to verify its reliability, we deployed a 1,000-machine Spark cluster and stress-tested it for three months. The stress test helped us identify a few bugs in the system, mostly in system memory management that caused the Spark nodes to crash. After fixing these bugs, the system ran smoothly for several weeks with very few crashes, this confirmed our belief that Spark could be a viable solution for distributed computing platform for autonomous driving.

Second, to quantify performance, we ran a high number of production SQL queries on MapReduce and on a Spark cluster. With the same amount of computing resources, Spark outperformed MapReduce by 5X on average. Using an internal query that we performed daily at Baidu, it took MapReduce more than 1,000 s to complete, but it only took Spark 150 s to complete.

9.2.2   DISTRIBUTED STORAGE

After selecting a distributed computing engine, we needed to decide on the distributed storage engine. Again, we faced two options, to remain with the Hadoop Distributed File System (HDFS) [11], which provides reliable persistent storage, or to use Alluxio, a memory-centric distributed storage system, enabling reliable data sharing at memory-speed, across cluster frameworks [3].

Specifically, Alluxio utilizes memory as the default storage medium and delivers memory-speed read and write performance. However, memory is a scarce resource and thus Alluxio may not provide enough storage space to store all the data.

The space requirement can be fulfilled by Alluxio's tiered storage feature. With tiered storage, Alluxio can manage multiple storage layers including Memory, SSD, and HDD. Using tiered storage, Alluxio can store more data in the system at the same time, since memory capacity may be limited in some deployments. Alluxio automatically manages blocks between all the configured tiers, so users and administrators do not have to manually manage the locations of the data. In a way, the memory layer of the tiered storage serves as the top-level cache, SSD serves as the second level cache, HDD serves as the third level cache, while persistent storage is the last level storage.

In our environment, we co-locate Alluxio with the compute nodes, and have Alluxio as a cache layer to exploit spatial locality. As a result, the compute nodes can read from and write to Alluxio; Alluxio then asynchronously persists data into the remote storage nodes. Using this technique, we managed to achieve a 30X speed up when compared to using HDFS only.

9.2.3 HETEROGENEOUS COMPUTING

By default, the Spark distributed computing framework uses a generic CPU as its computing substrate, which, however, may not be the best for certain type of workloads. For instance, GPUs inherently provide enormous data parallelism, highly suitable for high-density computations, such as convolutions on images. For instance, we have compared the performance of GPU vs. CPU on Convolution Neural Network-based object recognition tasks, and found that GPU can easily outperform CPU by a factor of 10–20 X. On the other hand, FPGA is a low-power solution for vector computation, which is usually the core of computer vision and deep learning tasks. Utilizing these heterogeneous computing substrates will greatly improve performance as well as energy efficiency.

There are several challenges on integrating these heterogeneous computing resources into our infrastructure: first, how to dynamically allocate different computing resources for different workloads. Second, how to seamlessly dispatch a workload to a computing substrate.

As shown in Figure 9.2, to address the first problem, we used YARN and Linux Container (LXC) for job scheduling and dispatch. YARN provides resource management and scheduling capabilities for distributed computing systems, allowing multiple jobs to share a cluster efficiently. LXC is an operating-system-level virtualization method for running multiple isolated Linux systems on the same host. LXC allows isolation, limitation, and prioritization of resources, including CPU, memory, block I/O, network, etc. Using LXC, one can effectively co-locate multiple virtual machines on the same host with very low overhead. Our experiments show that the CPU overhead of hosting a LXC is less than 5% comparing to running an application natively.

When a Spark application is launched, it can request heterogeneous computing resources through YARN. YARN then allocates LXCs to satisfy the request. Note that each Spark worker can host multiple containers, and that each may contain CPU, GPU, or FPGA computing resources. In this case, containers provide resource isolation to facilitate high resource utilization as well as task management.

To solve the second problem, we needed a mechanism to seamlessly connect the Spark infrastructure with these heterogeneous computing resources. Since Spark uses Java Virtual Machine (JVM) by default, the first challenge is to deploy workloads to the native space. As mentioned before, since the Spark programming interface centered on RDD, we developed a heterogeneous computing RDD which could dispatch computing tasks from the managed space to the native space through the Java Native Interface (JNI).

Next, in the native environment, we needed a mechanism to dispatch workloads to GPU or FPGA, for which we chose to use OpenCL due to its availability on different heterogeneous computing platforms. Functions executed on an OpenCL device are called kernels. OpenCL defines an API that allows programs running on the host to launch kernels on the heterogeneous devices and manage device memory.

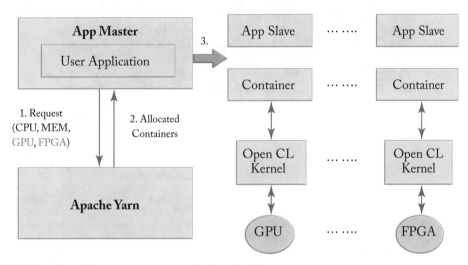

Figure 9.2: Distributed heterogeneous computing platform.

9.3   SIMULATION

With the unified infrastructure ready, let us now examine the services running on top of it. The first service we examine is distributed simulation tests for new algorithm deployment.

Whenever we develop a new algorithm, we need to test it thoroughly before we can deploy it on real cars, lest the testing cost is enormous and the turn-around time too high. Therefore, we can test the system on simulators [5]. One simulation approach consists in replaying the data through Robot Operating System (ROS) [6], where the newly developed algorithms are deployed for quick verification and early problem identification. Only after an algorithm passes all simulation tests can it be qualified to deploy on an actual car for on-road testing.

If we were to test the new algorithm on a single machine, it would either take too long or we would not have enough test coverage. To solve this problem, we leverage the Spark infrastructure to build a distributed simulation platform. This allows us to deploy the new algorithm on many compute nodes, feed each node with different chunks of data, and, at the end, aggregate the test results.

To seamlessly connect ROS and Spark, we needed to solve two problems: first, Spark by default consumes structured text data. However, for simulation we need Spark to consume multimedia binary data recorded by ROS such as raw or filtered readings from various sensors, detected obstacle bounding boxes from perception. Second, ROS needs to be launched in the native environment, where Spark lives in the managed environment.

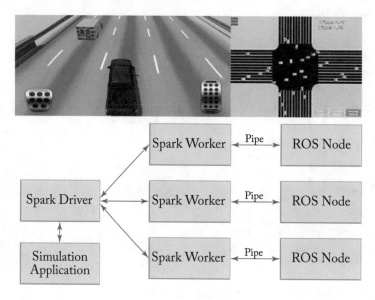

Figure 9.3: Simulation platform for autonomous driving.

9.3.1 BINPIPERDD

To make this architecture work, the first task is to have Spark consume binary input stream such as multimedia data. In the original design of Spark, inputs are in text format. Under such a context, we can have input records, as an example, with keys and values separated by space/tab characters, and records separated by Carriage Return characters. However, such an assumption is no longer valid in the context of binary data streams in which each data element in a key/value field could be of any value. To tackle this problem, we designed and implemented BinPipeRDD. Figure 9.4 shows how BinPipeRDD works in a Spark executor. First, the partitions of binary files go through encoding and serialization stages to form a binary byte stream. The encoding stage will encode all supported inputs format including strings (e.g., file name) and integers (e.g., binary content size) into our uniform format, which is based on byte array. Afterward, the serialization stage will combine all bytes arrays (each may correspond to one input binary file) into one single binary stream. Then, the user program, upon receiving that binary stream, would de-serialize and decode it according to interpret the byte stream into an understandable format. Next, the user program would perform the target computation (User Logic), which ranges from simple tasks such as rotate the jpg file by 90° if needed, to relatively complex tasks such as detecting pedestrians given the binary sensor readings from LiDAR scanners. The output would then be encoded and serialized before being passed in the form of RDD [Bytes] partitions. In the last stage, the partitions can be returned to the Spark driver through a collect operation or be stored in HDFS as binary files. With this process, we can now process and transform binary data into a user-defined format and transform the output of the Spark computation into a byte stream for collect operations or take it one step further to convert the byte stream into text or generic binary files in HDFS according to the needs and logic of applications.

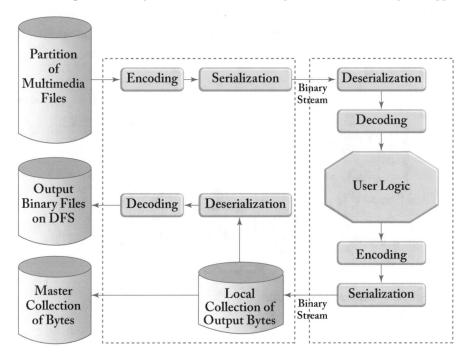

Figure 9.4: BinPipeRDD design.

9.3.2 CONNECTING SPARK AND ROS

With BinPipeRDD, Spark can now consume ROS Bag data, and we needed a way to launch ROS nodes in Spark as well as a way to communicate between Spark and ROS nodes. One choice was to design a new form of RDD to integrate ROS nodes and Spark, but this might involve changing ROS's as well as Spark's interfaces. Worrying about maintaining different versions of ROS, we went for a different solution and launched ROS and Spark independently, while co-locating the ROS nodes and Spark executors, and having Spark communicate with ROS nodes through Linux pipes. Linux pipes create a unidirectional data channel that can be used for inter-process communication. Data written to the write end of the pipe is buffered by the kernel until it is read from the read end of the pipe.

9.3.3   PERFORMANCE

As we developed the system, we continually evaluated its performance. First, we performed basic image feature extraction tasks on one million images (total dataset size > 12 TB) and tested the system's scalability. As shown in Figure 9.5, as we scaled from 2,000 CPU cores to 10,000, the execution time dropped from 130 s to about 32 s, demonstrating extremely promising capability of linear scalability. Next we ran an internal replay simulation test set. On a single node, it takes about 3 hr to finish the whole dataset. As we scale to eight Spark nodes, it only takes about 25 min to finish the simulation, again demonstrating excellent potential for scalability.

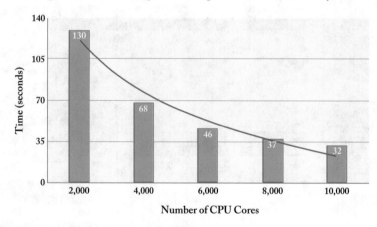

Figure 9.5: Simulation platform data pipeline.

9.4   MODEL TRAINING

The second application this infrastructure needs to support is offline model training. To achieve high performance in offline model training, our infrastructure provides seamless GPU acceleration as well as in-memory storage support of parameter servers.

As we use different deep learning models in autonomous driving, it is imperative to provide updates that will continuously improve the effectiveness and efficiency of these models. However, since the amount of raw data generated is enormous, we would not be able to achieve fast model training using single servers. To approach this problem, we developed a highly scalable distributed deep learning system using Spark and Paddle [10]. In the Spark driver, we can manage a Spark context and a Paddle context, and in each node, the Spark executor hosts a Paddler trainer instance. On top of that, we can use Alluxio as a parameter server for this system. Using this system, we have achieved linear performance scaling, even as we add more resources, proving that the system is highly scalable.

9.4.1   WHY USE SPARK?

The first question one may ask is why use Spark as the distributed computing framework for offline training, given that the existing deep learning frameworks all have distributed training capabilities. The main reason is that although model training looks like a standalone process, it may depend on the data preprocessing stage, such as ETL and simple feature extraction etc. As shown on the left side of Figure 9.6 below, in our practical tests, if we treated each stage as standalone, this would involve intensive I/O to the underlying storage, such as HDFS. As a consequence, we discovered that the I/O to the underlying storage often became the bottleneck of our whole processing pipeline.

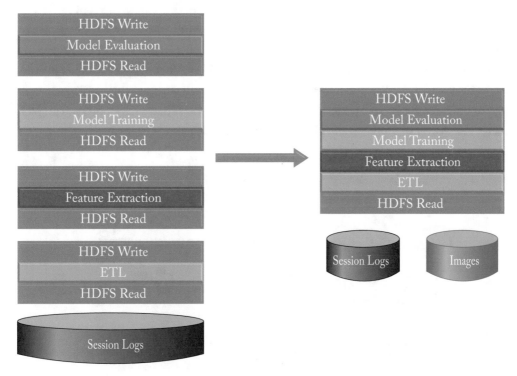

Figure 9.6: Training platform for autonomous driving.

As shown on the right side of Figure 9.6, by using Spark as the unified distributed computing framework, we can now buffer the intermediate data in memory, in the form of RDDs. The processing stages naturally form a pipeline without intensive remote IO accesses to the underlying storage in between the stages. This way, we read the raw data from HDFS at the beginning of the pipeline, and then pass the processed data to the next stage in the form of RDDs, until we finish the last stage and at last write the data back to HDFS. This approach allowed us to effectively double, on average, the throughput of the system.

9.4.2 TRAINING PLATFORM ARCHITECTURE

Figure 9.7 shows the architecture of our training platform. First, we have a Spark driver to manage all the Spark nodes, with each node hosts a Spark executor and a Paddle trainer, which allows us to utilize the Spark framework to handle distributed computing and resource allocation.

With this architecture, we can exploit data parallelism by partitioning all training data into shards so that each node independently processes one or more shards of the raw data. To synchronize the nodes, at the end of each training iteration, we need to summarize all the parameter updates from each node, perform calculations to derive a new set of parameters, and then broadcast the new set of parameters to each node so they can start the next iteration of training.

It is the role of the parameter server to efficiently store and update the parameters. If we were to store the parameters in HDFS, then again, as we have alluded to earlier, I/O would become the performance bottleneck. To alleviate this problem, we utilized Alluxio as our parameter server. As shown in Section 9.2.2, Alluxio is a memory-centric distributed storage, which utilizes in-memory storage to optimize for its I/O performance. Comparing to HDFS, we have observed an I/O performance gain factor of more than 5X by utilizing Alluxio as parameter servers.

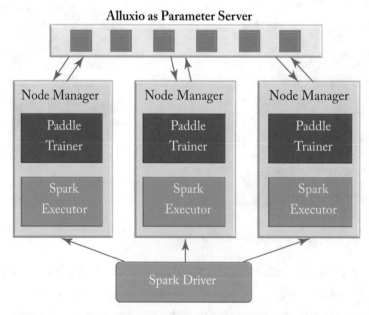

Figure 9.7: Training platform for autonomous driving.

9.4.3   HETEROGENEOUS COMPUTING

Next, we explored how heterogeneous computing could improve the efficiency of offline model training. As a first step, we explored how GPU performed compared to a CPU with Convolution Neural Networks (CNN). Using an internal object recognition model with the OpenCL infrastructure presented in Section 9.2.3, we have observed a 15X speed-up using GPU. The second step was to understand the scalability of this infrastructure. On our machine, each node is equipped with one GPU card. Figure 9.8 shows the result of this study, as we scaled the number of GPUs, the training latency per pass dropped almost linearly. This result confirmed the scalability of our platform, such that as we have more data to train against, we could reduce the training time by providing it with more computing resources.

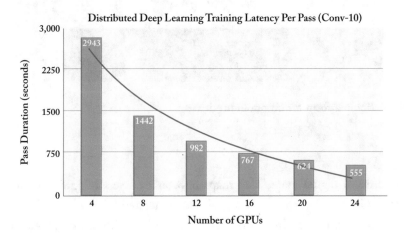

Figure 9.8: Performance of distributed model training.

9.5   HD MAP GENERATION

The third application this infrastructure needs to support is HD map generation, a multi-stage pipeline. By using Spark and heterogeneous computing, we managed to reduce the IO between the pipeline stages and accelerate the critical path of the pipeline.

As shown in Figure 9.9, like offline training, HD map production is also a complex process that involves many stages, including raw data reading, filtering and preprocessing, pose recovery and refinement, point cloud alignment, 2D reflectance map generation, HD map labeling, as well as the final map outputs [7, 8]. Using Spark, we can connect all these stages together in one Spark job. A great advantage is that Spark provides an in-memory computing mechanism, such that we do not have to store the intermediate data in hard disk, thus greatly reducing the performance of the map production process.

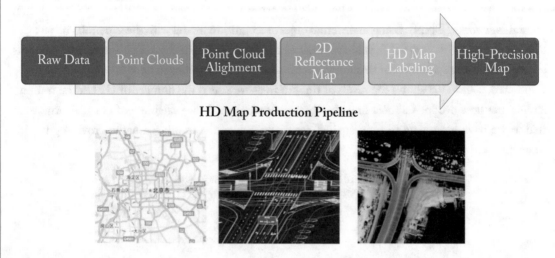

Figure 9.9: Simulation platform for autonomous driving.

9.5.1 HD MAP

Just as with traditional digital maps, HD maps have many layers of information. As shown in Figure 9.10, at the bottom layer we have a grid map generated by raw LiDAR data, with a grid granularity of about 5 cm by 5. This grid basically records elevation and reflection information of the environment in each grid cell. As the autonomous vehicles are moving and collecting new LiDAR scans, they compare in real time the new LiDAR scans against the grid map with initial position estimates provided by GPS and/or IMU, which then assists these vehicles in precisely self-localizing in real-time.

On top of the grid layer, there are several layers of semantic information. For instance, the reference line and lane information are added to the grid map to label each lane. This allows autonomous vehicles to determine whether they are on the correct lane when moving, and to also decide whether they are maintaining a safe distance to the vehicles on neighboring lanes. On top of the lane information, traffic sign labels will be added to notify the autonomous vehicles of the current speed limit, and whether traffic lights are nearby, etc. This gives an additional layer of protection in case the sensors on the autonomous vehicles fail to catch the signs.

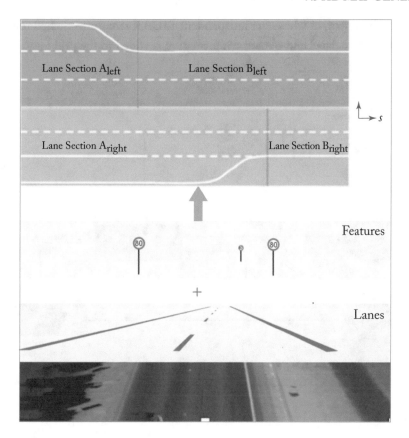

Figure 9.10: Performance of distributed model training.

9.5.2   MAP GENERATION IN THE CLOUD

Although we mentioned the importance of LiDAR data in HD map generation, it is not the only sensor data used. As shown in Figure 9.11, the HD map generation process actually fuses raw data from multiple sensors in order to derive accurate position information. First, the wheel odometry data and the IMU data can be used to perform propagation, or to derive the displacement of the vehicle within a fixed amount of time. Then the GPS data and the LiDAR data can be used to correct the propagation results in order to minimize errors.

In terms of process, the computation of map generation can be divided into three stages: first, Simultaneous Localization And Mapping (SLAM) is performed to derive the location of the each LiDAR scan. In this stage, the Spark job loads all the raw data, including IMU log, wheel odometry log, GPS log, and LiDAR raw data from HDFS. Second, it performs map generation and point cloud alignment, in which the independent LiDAR scans are stitched together to form a continuous map. Third, label and semantic information is added to the grid map.

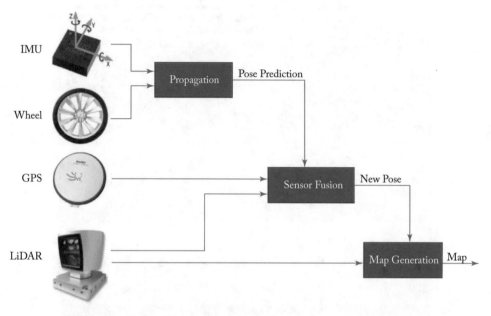

Figure 9.11: Map generation in the cloud.

Just as with offline training applications, we linked these stages together using a Spark job and buffered the intermediate data in memory. By using this approach, we achieved a 5X speedup when compared to having separate jobs for each stage. Also, the most expensive operation for the map generation stage is the iterative closest point (ICP) point cloud alignment [12]. By using the heterogeneous infrastructure, we managed to accelerate this stage by 30X by offloading the core of ICP operations to GPU.

9.6   CONCLUSIONS

An autonomous driving cloud is an essential part of the autonomous driving technology stack. In this chapter, we have shown the details of our practical experiences of building a production autonomous driving cloud. To support different cloud applications, we need an infrastructure to provide distributed computing, distributed storage, as well as hardware acceleration through heterogeneous computing capabilities.

If we were to tailor the infrastructure for each application, we would have to maintain multiple infrastructures, potentially leading to low resource utilization, low performance, and high management overhead. We solved this problem by building a unified infrastructure with Spark for distributed computing, Alluxio for distributed storage, and OpenCL to exploit heterogeneous computing resources for further performance improvement and energy efficiency.

With a unified infrastructure, many applications can be supported, including but not limited to distributed simulation tests for new algorithm deployment, offline deep learning model training, and HD map generation. We have delved into each of these applications to explain how the infrastructure can be utilized to support the specific features, and to provide performance improvement as well as scalability.

At this point, we are in the early stages of the development of a cloud infrastructure for autonomous vehicles, as autonomous driving technologies are actively evolving. Nonetheless, we know that, by having a unified infrastructure to provide the basic capabilities, including distributed computing, distributed storage, and heterogeneous computing, autonomous driving cloud itself can quickly evolve to meet the needs of emerging autonomous driving cloud applications.

9.7   REFERENCES

[1]   Liu, S., Peng, J., and Gaudiot, J.L. 2017. Computer, drive my car! *Computer*, 50(1), pp. 8–8. DOI: 10.1109/MC.2017.2. 173

[2]   Zaharia, M., Chowdhury, M., Franklin, M.J., Shenker, S., and Stoica, I. 2010. Spark: Cluster computing with working sets. *HotCloud*, 10(10-10), p. 95. 174, 175

[3]   Li, H., Ghodsi, A., Zaharia, M., Shenker, S., and Stoica, I. 2014. Tachyon: Reliable, memory speed storage for cluster computing frameworks. In *Proceedings of the ACM Symposium on Cloud Computing* (pp. 1-15). ACM. DOI: 10.1145/2670979.2670985. 174, 175

[4]   Stone, J.E., Gohara, D., and Shi, G. 2010. OpenCL: A parallel programming standard for heterogeneous computing systems. *Computing in Science & Engineering*, 12(3), pp. 66–73. DOI: 10.1109/MCSE.2010.69. 174

[5]   Basarke, C., Berger, C., and Rumpe, B. 2007. Software & systems engineering process and tools for the development of autonomous driving intelligence. *Journal of Aerospace Computing, Information, and Communication*, 4(12), pp. 1158–1174. DOI: 10.2514/1.33453. 177

[6]   Robot Operating System, http://www.ros.org/. 177

[7]   Levinson, J. and Thrun, S. 2010. Robust vehicle localization in urban environments using probabilistic maps. In *2010 IEEE International Conference on Robotics and Automation (ICRA)*, (pp. 4372–4378). IEEE. DOI: 10.1109/ROBOT.2010.5509700. 183

[8]   Schreiber, M., Knöppel, C., and Franke, U. 2013. Laneloc: Lane marking based localization using highly accurate maps. In *Intelligent Vehicles Symposium (IV), 2013 IEEE* (pp. 449–454). IEEE. DOI: 10.1109/IVS.2013.6629509. 183

[9]   Geiger, A., Lenz, P., Stiller, C., and Urtasun, R. 2013. Vision meets robotics: The KITTI dataset. *The International Journal of Robotics Research*, 32(11), pp. 1231–1237. DOI: 10.1177/0278364913491297.

[10]   Baidu PaddlePaddle: https://github.com/PaddlePaddle/Paddle. 180

[11]   White, T. 2012. *Hadoop: The Definitive Guide*. O'Reilly Media, Inc. 175

[12]   Segal, A., Haehnel, D., and Thrun, S. 2009. Generalized-ICP. In *Robotics: Science and Systems*. DOI: 10.15607/RSS.2009.V.021. 186

CHAPTER 10

Autonomous Last-Mile Delivery Vehicles in Complex Traffic Environments

10.1 BACKGROUND AND MOTIVATIONS

In this chapter, we present a case study of a commercial autonomous last-mile delivery vehicle operating in complex traffic environments.

Logistics service is an essential component of a satisfactory e-commerce experience because consumers commonly expect the purchased goods to be delivered safely and quickly. Last-mile delivery here refers to the delivery of goods ordered online from a local distribution center to the consumers. As the final stage of the e-commerce delivery process, autonomous last-mile delivery is extremely challenging as it must handle complex traffic environments.

The main motivations for autonomous last-mile delivery services lies in the intrinsic disadvantages suffered by traditional last-mile deliveries: first, growing labor cost may be prohibitive for service providers like JD.com, a leading e-commerce company in China. Indeed, from our operational data, a contracted delivery clerk with an annual salary of almost $20,000 can deliver 110 e-commerce parcels per day, which renders that each delivery order costs nearly 0.5 USD. This cost is expected to continue increasing as the demographic dividend has reached its end. Secondly, a delivery clerk has to spend time on repeatedly contacting the consumers, waiting for their pick-ups, and traveling on the road, where time is wasted which diverts humans from other more creative work. Fortunately, autonomous driving technologies respond to this exact problem and the benefits of utilizing an autonomous driving vehicle to replace a delivery clerk lie in the following aspects.

1. The delivery is no longer interrupted by weather conditions or time. Ideally, an unmanned vehicle can respond in a timely manner to the consumers' 24/7 e-commerce orders. Particularly, delivering at night or during legal holidays should not incur extra costs.

2. Similarly, since an unmanned vehicle is designed to operate 24/7, it has more temporal flexibility than the conventional 8-hr working time to accomplish the delivery tasks.

3. The costs in terms of labor recruitment, training, and management are completely removed.

4. An autonomous vehicle system improves the safety and efficiency of both delivery clerks and other people sharing the public transportation infrastructure.

Therefore, deploying autonomous vehicles for the last-mile delivery is a promising approach to overcome the aforementioned disadvantages.

10.2   AUTONOMOUS DELIVERY TECHNOLOGIES IN COMPLEX TRAFFIC CONDITIONS

Autonomous driving technologies have been extensively studied in the past few years [1]. However, the traffic conditions in unruly environments are more challenging, primarily because of the large numbers of heterogeneous traffic participants, including pedestrians, bicycles, and automobiles, sharing the roads, and these traffic participants do not necessarily follow traffic rules. These unruly environments are common in various metropolis of China, which is distinct from most of the cases in more developed countries, e.g., North America or Europe.

1. Since the urban population is large in China, urban residents typically live in apartments rather than individual houses. This naturally means that the population density is high around apartment complexes, which is quite distinct from the situations in more developed countries, which have most of their population in the suburbs. As a result, when traveling in the unstructured environment around apartment environments, an autonomous vehicle would commonly encounter large numbers of complicated interactive objects such as automatic barrier gates in parking lots, pedestrians, bicycle riders, etc. (Figure 10.1a).

2. There are typically multiple types of traffic participants including bicycles, electric bicycles, motorcycles on urban roads, etc. Each traffic participant has its own kinematic feature (Figure 10.1b). To make the situation worse, some residents may use their vehicles in unusual, unsafe, or even illegal ways (e.g., the man riding a bicycle while holding an unusually long object in Figure 10.1c, which is actually not a rare scene in areas with dense populations).

3. As depicted in Figure 10.1d, traffic jams often happen in large cities. This is not surprising because motor vehicle ownership keeps growing rapidly.

Figure 10.1: Typical scenarios that reflect the complexity of unruly driving behaviors: (a) an un-structured parking lot that various traffic tools share the space; (b) urban road with multiple types of two-wheel vehicles; (c) a man who rides carrying with long sticks; and (d) traffic congestion. These factors show that the difference of traffic conditions between more or less developd countries coun-tries(e.g., China).

Since the traffic conditions in unruly environments (e.g., China) are far more complex than those in developed countries, existing autonomous driving technologies suitable for developed countries might not be directly applicable in a chaotic environment. They should be enhanced and adapted to the application scenarios and budget constraints.

Safety should obviously be a primary concern in developing and maintaining autonomous vehicles. Compared with autonomous vehicles that carry passengers, delivery vehicles have unique safety requirements. They should at the same time obey all traffic laws and make the traffic on the road proceed in a normal and safe manner. If this is not possible, the delivery autonomous vehicles should not cause an impediment or a danger to the other vehicles, especially other vechicles car-rying human passengers. Instead, a delivery vehicle might have to sacrifice itself as an alternative choice. This means a special safety requirements for autonomous delivery vehicles. This design requirement will be discussed in detail in later sections.

10.3   JD.COM: AN AUTONOMOUS DRIVING SOLUTION

As one of the largest e-commerce companies, JD.com develops autonomous vehicles for last-mile delivery mainly to reduce delivery costs. The cost to deliver each e-commerce parcel would be reduced by 22.45% if autonomous vehicles were used. This conclusion was made under the assumption that each vehicle can deliver 60 e-commerce parcels per day. If 10% of the entire e-commerce orders in JD.com are delivered by autonomous vehicles, this would result in a cost saving of at least $110 million annually. More generally speaking, if JD.com can be said to deal with 5% of the e-commerce parcel delivery orders with autonomous driving technologies in the entire market, it would reduce annual costs by a whopping $7.64 billion. The promising benefits and the rational profit mode have been the motivation to carry on the R&D of autonomous vehicles for the last-mile delivery.

10.3.1   AUTONOMOUS DRIVING ARCHITECTURE

As illustrated in Figure 10. 2, it takes more than 20 modules working simultaneously and cooperatively to make an autonomous driving system work.

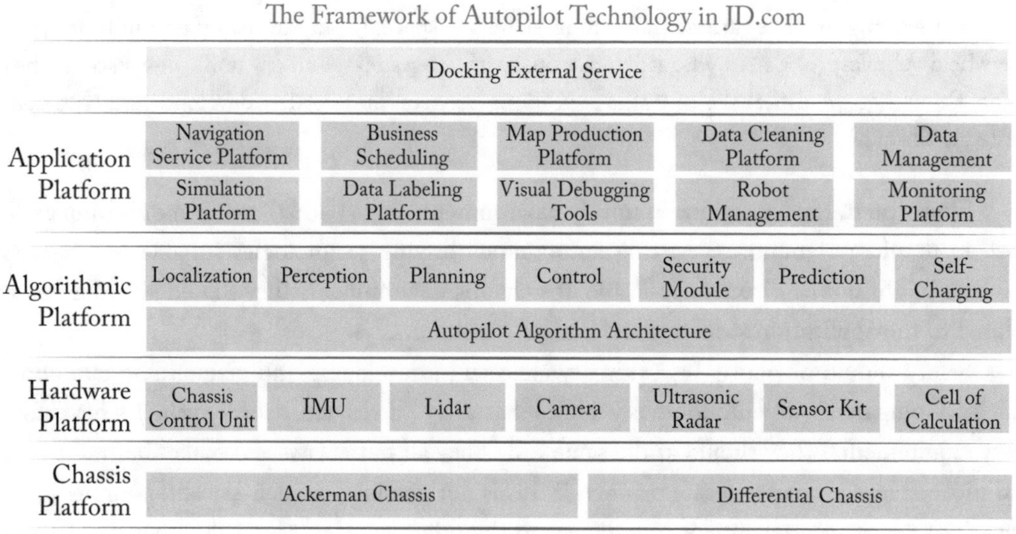

Figure 10.2: The architecture of our autonomous driving system.

Broadly speaking, each of the modules shown in Figure 10. 2 is either an online or an offline module. An online module functions when an autonomous vehicle travels on the road [2], whereas an offline module mainly functions for the purpose of offline feature extraction, training, configuration, simulation, testing, and/or evaluation [3].

In a last-mile delivery scheme, a delivery address is assigned to the autonomous vehicle. Thereafter, the current location of the vehicle, as well as the route from the current position to the destination are specified through the navigation service platform. The vehicle then begins to move. During its trip toward the destination, the autonomous vehicle implements localization, local perception, local prediction, local decision, local planning, control, etc. in the algorithmic platform to guarantee that the trip is safe, smooth, efficient, and predictable.

The hardware platform is a supporting platform for the functioning of the aforementioned elements. The hardware platform consists of the concrete devices, as well as the connections and managements among them. The offline modules, on the other hand, prepare for the aforementioned online modules. For example, the map production platform is responsible for the creation of the High-Definition (HD) map; the simulation platform and visual debugging tools help the developers debug. In the remaindering of this section, we would like to introduce a few highlighted modules in the autonomous driving technology stack.

10.3.2 LOCALIZATION AND HD MAP

Localization is responsible for deriving the current location of the vehicle. In our autonomous driving solution, there are broadly five sources of information which contribute to localization. The first one is the GPS signal, which is used only once as a startup signal to activate the request of an HD map for later usage

The remaining four localization sub-modules include localization algorithms based on multi-line Lidar, cameras, chassis-based odometry, and Inertial Measurement Unit (IMU) Information. In using a multi-line Lidar algorithm, the generated point clouds are matched to the pre-stored ones in the requested HD map. The matching is implemented via the generalized Iterative Closest Point (ICP) method [4] online, where the Lidar point clouds are pre-separated into ground and non-ground point clouds.

With multi-line LiDAR, camera, and chassis data at hand, the derivative of the data indicates the movement, which is odometry. Through synthesizing the independent sources of odometry information, a reliable odometry result is derived, otherwise known as as odometry fusion. The fused result, when synthesized with IMU results, is utilized to calibrate the ICP method in a Kalman filter framework [5]. In turn, the ICP method could be further used to calibrate the fusion result so that they could be both closer to the ground truth. Such an intricate architecture is illustrated in Figure 10.3.

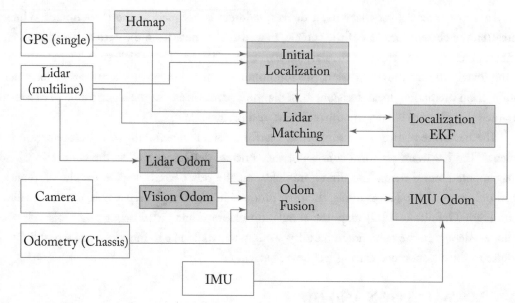

Figure 10.3: The architecture of localization.

Our HD map contains the detailed properties of the road, which is represented by 16 layers to provide both static and dynamic information of the environment. They are composed of multiple sub-maps including a geometric map, semantic map, and real-time map. Some of the elements shared by both autonomous passenger vehicles and delivery vehicles include, for example, the location, width, type, curvature, and boundaries of each lane and road, the intersections with their associated semantic features like crosswalks, traffic signals, speed bumps, etc. Meanwhile, road elements specifically for delivery vehicles in China are included. For example, detailed representation of pillars (commonly placed on the entrance of bicycle lanes to prohibit large vehicles' access), clear areas (restricted areas where the vehicle cannot stop, could be temporary), gates (gates to access distribution stations or to enter residential communities), and safety islands (areas inside a large intersection where pedestrains/cyclists can wait for the next green light) are important in our map. Beside lanes and roads that commonly exist in other HD maps, we added "lane group" elements as an intermediate level between lane and road for better lane associations at large intersections.

Delivery vehicles require cost effective construction of HD maps with the same sensors used on the vehicle, and at all districts in the urban areas, where the GPS signals are usually weak. They also require frequent updates to HD maps since the roads in urban areas in China change more often than in other places. This requires the establishment of a technical team that focuses on building and maintaining the in-house HD maps based on modern sensor fusion and SLAM technologies that do not rely on a differential GPS or expensive IMUs. Machine learning techniques developed by the perception team are incorporated to assist detection of static vehicles and traffic lights to improve the efficiency of map construction.

10.3.3 PERCEPTION

The perception module is responsible for recognizing and tracking the dynamics of the obstacles in the environment. The perception module plays a key role in making the whole autonomous system capable of running through the complex traffic environments. The hardware setup, as illustrated in Figure 10.4, is a common multi-source solution. Our vehicle is equipped with one 1-beam LiDAR, one 16-beam LiDAR, four mono cameras to detect the surrounding objects. A high-resolution HDR camera is used for traffic light detection. Ultrasonic receptors are used to prevent immediate collisions with other objects or road elements.

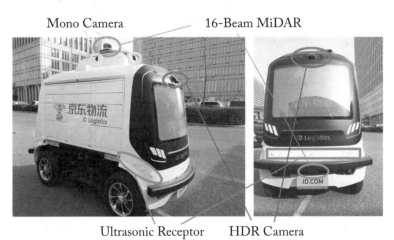

Figure 10.4: Sensor hardware setup.

As illustrated in Figure 10.5, in object detection, three methods are independently applied, and their detection results are fused. Concretely, the first two detectors process the point cloud data using machine-learning and geometry-based methods, respectively. The third detector processes the visual data using the machine-learning-based methods.

From our practical experience, we observed that the first method is good at generic traffic-related object classification, but the performance is poor at dealing with pedestrians, which is an inherent drawback of the learning-based methods, because the features of pedestrians are not easy to detect compared to traffic signs. Conversely, the second method provides stable results in recognizing objects with typical shapes (e.g., the pedestrians). The third method is particularly suitable for recognizing partially observed objects because it focuses on color, shading, and texture as identifying features. The three detectors complement each other to provide comprehensive and reliable object recognition results.

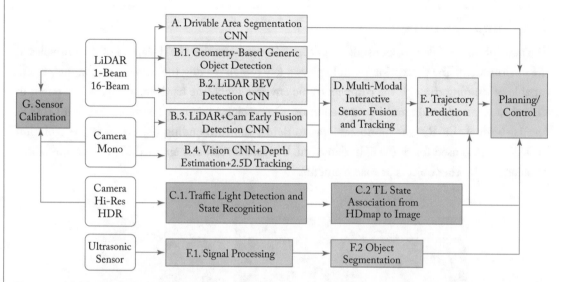

Figure 10.5: The architecture of perception.

As discussed in Section 10.1, many types of on-road elements exist on the roads and affect autonomous vehicles. Taking the scenario in Figure 10.1c as an example, the first time our autonomous vehicle encounters such illegal traffic, since the algortithm has not encountered so much data as to correctly realize the relationship between the electric bicycle and the long stick, the ego vehicle could be stuck on the road and not know how to handle this scenario. When observing such a case, a remote monitoring human may take over the vehicle as a precaution. Later, this object is placed into a newly established category, and some existing data are marked manually for learning this category in both detectors 1 and 3. We believe that supervised learning-based enumeration is the only feasible way toward achieving reliable recognition performance. In addition, a deduction approach is developed to systematically estimate the status of the interested but unobservable surrounding traffic, such as the status of the traffic lights and that of the vehicle in front of the vehicle right ahead.

10.3.4 PREDICTION, DECISION, AND PLANNING

The prediction, decision, and planning module is critical to enabling our autonomous vehicles to safely navigate through complex traffic conditions. Prediction refers to approximating the future trajectories of the tracked moving obstacles. Prediction is done in two cooperative layers. Concretely in the first layer, the behaviors of the well-tracked vehicles are predicted with the utilization of the routing information and lane information in the HD map. Herein, we build assumptions on how well the tracked objects will abide by the traffic rules based on the historic road data on this road. The first layer is commonly suitable to handle the regulated vehicles. The second layer targets the prediction of abnormal object behavior, which is achieved through machine learning-based methods and deduction methods. The two layers work together to provide the trajectories of the moving objects of interest in the sighting scale of the vehicle. The predicted trajectory is also attached with a variance to measure the prediction confidence.

Decision and planning play a critical role in directly guiding the local driving maneuvers of the vehicle. In dealing with the complex scenarios in a chaotic environment, the decision module should find a rough but reliable homotopic trajectory, while the planning module should run fast. In the decision module, a sample-and-search-based method is adopted. In more details, while searching in a graph consisting of sampled nodes, a dynamic programming algorithm is used, and the cost function considers the uncertainties (quantified by confidence degrees mentioned above) in the perception and prediction modules. Through this, a robust and safe trajectory decision is made.

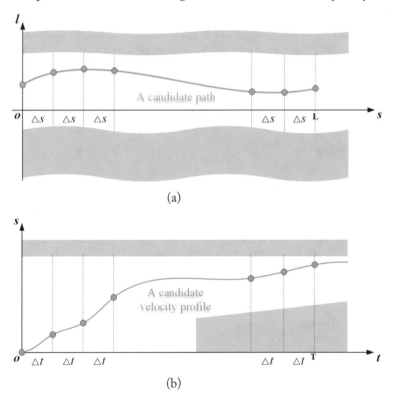

(a)

(b)

Figure 10.6: Schematics on the quadratic programming model formulation: (a) path planning in an *s-l* coordinate system; and (b) velocity planning in an *s-t* coordinate system.

The coarse trajectory derived from the decision module naturally decides whether to yield, or to bypass each of the obstacles. With this coarse trajectory at hand, we can significantly reduce the solution space to a small neighborhood around that coarse trajectory to avoid wasted time in global planning. More specifically, the trajectory planning scheme is split into multiple phases, including decision generation, path planning, and velocity planning. Through this decomposition, the original difficulties are largely eased. In the decision generation phase, the interaction between the vehicle and the surrounding objects are determined in a "path" in a path-time coordinate system. Herein, the feasible bounds are inherently forming a tunnel which is determined by the coarse trajectory in the decision phase.

We will discuss this problem in a general format with an x-y coordinate system. It can be simply converted to path optimization using a frenet-frame l-s (Figure 10.6a) or speed optimization using a path-time frame s-t (Figure 10.6b). In more detail, assume the problem horizon along the x axis is fixed as X, we define as many as N straight lines which intersect the x axis at (N + 1) equidistant points from 0 to X. With this observation, the path and speed-planning schemes can each be converted so as to find the positions of N grids along the lines, so that connecting the grids in a sequence would form a candidate s-l path profile or an s-t velocity profile. Let us define the position of each grid as y_i ($i = 1,...,$N) and the derivative, second-order derivative, and third-order derivative of y_i are

$$y_i' \equiv {}^{dy_i}\!/dx$$

$$y_i'' \equiv {}^{dy_i'}\!/dx \tag{1}$$

$$y_i''' \equiv {}^{dy_i''}\!/dx.$$

We restrict the magnitude of each term to be within some bounds, as shown below. Notice that the y_i', y_i'', and y_i''' terms all could appear in the optimization targets. Limiting their magnitudes could prevent a dangerous turn or sudden jerk of the vehicle:

$$|y_i'| \leq y_{max}'$$

$$|y_i''| \leq y_{max}'' \tag{2}$$

$$|y_i'''| \leq y_{max}''',$$

where y_{max}', y_{max}'', and y_{max}''' are bounds for the corresponding profiles.

In addition, the two-point boundary conditions are restricted via a few boundary equalities at $x = 0$ and $x = $ X. For example, in speed optimization, $y_1(0)$, its derivative and second derivative are specified as the initial path position, initial velocity, and initial acceleration at $t = 0$. At $x = $ X,

$y_N(X)$ and its derivatives are set to 0 in path optimization or only the second-order and third-order derivatives of $yN(X)$ are set to 0 in the velocity optimization.

The vehicle should obviously not collide with the barriers of the tunnel. The configuration space is therefore formed under basic collision-avoidance constraints. In path optimization, they are formulated by covering the rectangular vehicle as a pair of circles to simplify the calculation [8], and then require each circle to be collision-free from the two barriers of the tunnel, while for speed optimization, the ego vehicle should not collide with any of the *s-t* regions in the path-time graph. Besides the aforementioned constraints, we expect the trajectory to be smooth, and the optimized result to be as close to the coarse trajectory as it possible. There are other optimization objects and constraints that are related to decisions that the results need to consider. This expectation is reflected in the following minimization objective:

$$J = \sum_k \sum_i \sum_{j=0,1,2,3} w_{k,i,j} \cdot (y_i^{(j)} - ref_{k,i,j})^2. \tag{3}$$

In Equation (3), $k \in \{1,...,N_{ref}\}$ refers to the index of reference profiles. Suppose there are Nref reference profiles which are simultaneously affecting the optimization performance, $i \in \{1,...,N\}$ denotes the index of grids. $w_{k,i,j} \geq 0$ ($j = 0,...,3$) denotes the weighting parameter to encourage the attraction toward the *k*th reference profile at the *i*th grid w.r.t. The *j*th order of derivative of y_i and $ref_{k,i,j}$ ($j = 0,...,3$) stands for the corresponding reference profile. As to why we have multiple $ref_{k,i,j}$ at each grid, let us take the velocity planning as an example, at the Nth grid, *n*. No one knows *a priori* how to specify the *s'* at the terminal moment (i.e., $y_N'(N)$). Thus, multiple possibilities are integrated in the form of a weighted sum.

Now with the optimization objective and the constraints at hand a Quadratic programming (QP) problems are formulated, wherein the cost function is quadratic and the constraints are all linear:

Minimize (3),
Subjected to
    Dynamic constraints (1),   (4)
    Magnitude constraints (2),
    Two-point boundary conditions, and
    Collision-avoidance constraints.

The nominal decision variables should be $\lfloor y_i'''$, and the degree of freedom of the path planning problem should be N (roughly). In Equation (4), nonetheless, we set the decision variables as y, y', y'', and y''' to achieve a higher numerical optimization speed (the rationale behind this can be found by our analysis of simultaneous strategy in [7]). The QP problem is then numerically solved via a local optimizer. The 99th percentile time consumption to solve the aforementioned QP problem is within 10 msec, thus the online planning is sufficient to react to the sudden appearance of events and/or of merged objects in complex road scenarios.

10.4   SAFETY AND SECURITY STRATEGIES

As mentioned in previous sections, safety is of paramount importance in designing autonomous delivery vehicles. Safety guarantees are established in multiple layers, including simulation-level, vehicle-end, and remote monitoring. In this section, we introduce a few highlighted safety strategies in our autonomous driving technology stack.

10.4.1 SIMULATION-LEVEL VERIFICATION

All submitted codes must be exercised through a large number of benchmark tests. In each test, the input raw data are loaded to "decorate" a virtual real-world test, and the performance of the autonomous vehicle is measured by the virtual output actions in the simulator with a number of carefully defined criteria. In creating the simulation test cases, we utilize the recorded real-world data to build tens of thousands of virtual scenarios. The codes are tested in those virtual but realistic scenarios for performance evaluation. The codes that can pass the simulation tests are tested further in the vehicle-end system, and then the newly emerged issues are recorded and manually marked as virtual scenarios for future code examination. Through this closed-loop approach, the safety of the development is rapidly improved.

10.4.2 VEHICLE-END MONITORING

We have implemented a vehicle-end low-level guardian module which monitors the occurrence or near-occurrence of an emergency. The guardian module primarily monitors the health of the control system and deals with internal and external exceptions from within the system and from outside. To address the sudden failures that may happen to the hardware, redundant units are equipped to the autonomous vehicles, in association with a failure detection monitor. If even the redundant units fail as well (for example, due to low temperatures that makes the visual sensors fail), the guardian module takes over as per the pre-defined failure-recovery rules. Regarding the troubles from the external environment, when obstacles are overly close to the vehicle, or approaching the vehicle at high speeds, the vehicle would take actions to reduce the collision risks.

10.4.3 REMOTE MONITORING

To guarantee the operational safety of our delivery vehicles, we have also developed a remote monitoring platform. The driving behavior of the vehicle are monitored in real time. The engineer who remotely monitors the vehicle could take over the control to assist the vehicle to get out of an abnormal situation. If the remotely monitoring engineer is absent, the remote platform would generate a warning signal to inform the police regarding the situation.

10.5   PRODUCTION DEPLOYMENTS

On the way toward large-scale production deployment, we hold a progressive strategy to divide the entire technical difficulties into four stages. The first stage is about autonomously driving at low speeds with manual surveillance. Herein, surveillance stands for the action to add an extra level of safety that a human will assist to make sure the behavior of the vehicle is as expected and can switch the driving mode to manual operation whenever any potential risk occurs. The second stage is about autonomously driving at low speeds without manual surveillance. The third and fourth stages are autonomously driving in relatively high speeds with/without manual surveillance, respectively. It is easy to see that there are no huge gaps between the four stages. On the way to raise the vehicle's driving speed, progress is made adaptively: if the technical developments are successful, the increases in speed can be relatively large, and conversely.

In addition to the aforementioned progressive technical roadmap, the journal to achieve profitability is also designed to be progressive. When the technologies were far from being appropriate for the on-road trial operations, we focused on developing the low-level chassis, which can be commonly used in many robotic applications. This idea is beneficial for three reasons: (i) The technologies for indoor low-level autonomous moving can be seamlessly applied to warehousing logistics, which will improve the autonomous ability of the full logistic chain; (ii) commercialization and profitability of robotic technogies will arrive before the autonomous driving technologies are ready for full deployment; and (iii) even when the autonomous driving technologies are not mature, the developers should be aware how to develop products, so as to avoid keeping their minds distant from the terminal target, i.e., production.

Regarding the business logic, efforts are also progressively made so that the efficiency of the autonomous last-mile delivery can be maximized. We have been making the scheduling qualities in the e-commerce platform, the warehouse, and the distribution centers take more care about the time efficiency of the deliveries. Improvements in the aforementioned three aspects are made cooperatively and simultaneously to render a good delivery service. Up to this point, we have deployed more than 300 self-driving vehicles for trial operations in several provinces of China, with an accumulated 715,819 miles.

10.6   LESSONS LEARNED

In deploying this autonomous vehicle system, a few lessons have been learned, the first being that the algorithms should be explainable, which make their performance easy to estimate, predict, and thus acceptable for other users sharing the road. Having said this, we find that deep-learning based end-to-end solutions are not pratical at this stage, but machine-mearning-based methods are extensively used in each sub-module with clearly defined boundaries. Secondly, the routes of last-mile delivery vehicles are mostly fixed, so that we heavily rely on HD maps to record fine details along the route. The third lesson is that, during the real-world on-road tests, the pursuit for higher taken-over mile index is misleading because it may lead the developers to hide the risks or problems rather than to find and conquer them. In our viewpoint, accurately recognizing a risk and then requesting a manual taken-over is highly meaningful, which is a critical part of the entire safety guarantee system. The fourth lesson is that it makes sense to clearly separate the jobs that are suitable for humans and for an automated machine. After a long time of operations in trial, we have learned that it is feasible to allocate the vehicles to handle the complicated but scenarios that repeatedly appear and the human monitors can take over the vehicles when required. In addition, autonomous driving does not mean that human beings become useless, instead they can do innovative work highly related to maintaining an autonomous delivery system with autonomous vehicles.

10.7 REFERENCES

[1]   Liu, S., Li, L., Tang, J., Wu, S., and Gaudiot, J.L. 2017. Creating autonomous vehicle systems. *Synthesis Lectures on Computer Science*, 6(1), pp. i–186. DOI: 10.2200/S00787ED-1V01Y201707CSL009. 190

[2]   Liu, S., Tang, J., Zhang, Z., and Gaudiot, J.L. 2017. Computer architectures for autonomous driving. *Computer*, 50(8), pp.18–25. DOI: 10.1109/MC.2017.3001256. 192

[3]   Liu, S., Tang, J., Wang, C., Wang, Q., and Gaudiot, J.L. 2017. A unified cloud platform for autonomous driving. *Computer*, 50(12), pp.42–49. DOI: 10.1109/MC.2017.4451224. 192

[4]   Segal, A., Haehnel, D., and Thrun, S. 2009. Generalized-icp, in *Robotics: Science and Systems (RSS)*, 2009. DOI: 10.15607/RSS.2009.V.021. 193

[5]   Li, W. and Leung, H. (2003, October). Constrained unscented Kalman filter based fusion of GPS/INS/digital map for vehicle localization. In *Proceedings of the 2003 IEEE International Conference on Intelligent Transportation Systems*, (2) pp. 1362–1367. IEEE. 193

[6]   Ma, W. C., Tartavull, I., Bârsan, I. A., Wang, S., Bai, M., Mattyus, G., Homayounfar, N., Lakshmikanth, S. K., Pokrovsky, A., and Urtasun, R. (2019). Exploiting sparse semantic HD maps for self-driving vehicle localization. arXiv preprint arXiv:1908.03274. DOI: 10.1109/IROS40897.2019.8968122.

[7]   Li, B. and Shao, Z. (2015). A unified motion planning method for parking an autonomous vehicle in the presence of irregularly placed obstacles. *Knowledge-Based Systems*, 86, 11–20. DOI: 10.1016/j.knosys.2015.04.016. 199

[8]   Fan, H., Zhu, F., Liu, C., Zhang, L., Zhuang, L., Li, D., Zhu, W., Hu, J., Li, H., and Kong, Q. 2018. Baidu apollo em motion planner, arXiv preprint arXiv:1807.08048. 198

CHAPTER 11

PerceptIn's Autonomous Vehicles Lite

11.1 INTRODUCTION

In this chapter, we present a case study of PerceptIn's affordable autonomous vehicles for microtransit services. Unlike other L4 autonomous vehicles that cost hundreds of thousands of dollars each, PerceptIn started with a more practical and affordable approach: start small, go slow, and target microtransit scenarios that cover the 1–5 miles distance [1, 2].

The problem is that today's mobility-as-a-service ecosystem often does not do a good job covering intermediate distances, say a few miles. Hiring an Uber or Lyft for such short trips proves awkwardly expensive, and riding a scooter or bike more than a mile or so can be taxing to many people. So, getting yourself to a destination that is from 1–5 miles away can be a challenge. Yet such trips account for about half of the total passenger miles traveled.

Many of these intermediate-distance trips take place in environments with limited traffic, such as university campuses and industrial parks, where it is now both economically reasonable and technologically possible to deploy small, low-speed autonomous vehicles powered by electricity. PerceptIn, now has autonomous vehicles operating in United States, Europe, Japan, and China. Because these diminutive autonomous vehicles never exceed 20 miles (30 km) per hour and do not mix with high-speed traffic, they do not engender the same kind of safety concerns that arise with autonomous cars that travel on regular roads and highways. While autonomous driving is a complicated endeavor, the real challenge for PerceptIn was not about making a vehicle that can drive itself in such environments—the technology to do that is now well established—but rather about keeping costs down.

Given how expensive autonomous cars still are in the quantities that they are currently being produced—an experimental model can cost you in the neighborhood of U.S. $800,000—you might think it impossible to sell a self-driving vehicle of any kind for much less. Our experience over the past few years shows that, in fact, it is possible today to produce a self-driving passenger vehicle much more economically.

11.2 EXPENSIVE AUTONOMOUS DRIVING TECHNOLOGIES

It is well known that autonomous driving is not one single technology but rather a complex system integrating many technologies, including sensing, localization, perception, decision making, high-definition (HD) map creation, and system integration. In this section we provide a brief overview of these technologies, and demonstrate that the high costs of sensors, computing systems, and HD maps are the major barriers of autonomous driving deployment.

11.2.1 SENSING

The typical sensors used in autonomous driving include Global Navigation Satellite System (GNSS), Light Detection and Ranging (LiDAR), cameras, radar, and sonar.

- **GNSS receivers**, especially those with real-time kinematic (RTK) capabilities, help autonomous vehicles localize themselves by updating global positions with at least meter-level accuracy. A high-end GNSS receiver for autonomous driving could cost well over $10,000.

- **LiDAR** is normally used for the creation of HD maps, real-time localization, as well as obstacle avoidance. LIDAR works by bouncing a laser beam off of surfaces and measuring the reflection time to determine distance. A typical LiDAR unit such as used in autonomous vehicles covers a range of 150 m and samples over 1 million spatial points (with <x, y, z> coordinates) per second. Each point is associated with a reflectivity attribute which can be used to identify points between different frames. By comparing the displacements of the spatial points between two frames, we can derive the displacement of the vehicle. However, LiDAR units suffer from two problems: (1) they are extremely expensive (a high-end autonomous driving grade LiDAR could cost over $80,000); (2) they may not provide accurate measurements under bad weather conditions, such as heavy rain or fog.

- **Cameras** are mostly used for object recognition and tracking tasks, such as lane detection, traffic light detection, and pedestrian detection. Existing implementations usually mount multiple cameras around the vehicle to detect, recognize, and track objects. However, an important drawback of camera sensors is that the data they provide may not be reliable under bad weather conditions and that their sheer amount creates high computational demands. Note that these cameras usually run at 60 Hz, and, when combined, can generate over 1 GB of raw data per second.

- **Radar and Sonar:** The radar and sonar sub-systems are used as the last line of defense in obstacle avoidance. The data generated by radar and sonar shows the distance from the nearest object in front of the vehicle's path. Note that a major advantage of radar is that it works under all weather conditions. Sonar usually covers a range of 0–10 meters whereas radar covers a range of 3–150 m. Combined, these sensors cost less than $1,000.

11.2.2 LOCALIZATION

Localization is about *providing accurate vehicle position updates* in real time and several different techniques can be utilized to achieve it.

- The natural choice for localization is to use **GNSS** directly. Nonetheless, this cannot be the sole source for localization because of multipath problems, meaning that signals may bounce off of buildings, introducing noise and delays. Further, GNSS requires an unobstructed view of the sky. In closed environments such as tunnels, its accuracy thus significantly degrades.

- **LiDAR** is commonly used in localization as well. First the localization sub-system extracts point clouds through LiDAR scans, providing a "shape description" of the environment. Then, the localization sub-system compares a specific observed shape against the shapes in a confined region of the HD map to reduce uncertainty and track the position of the moving vehicle.

- **Cameras** can be used for localization as well; the technology used is called visual odometry. Visual odometry works by first extracting spatial points (with <x, y, z> coordinates and a feature descriptor that uniquely identifies a spatial point) through stereo vision, and then by comparing the locations of the detected spatial points between consecutive frames to deduce the movement of the vehicle between the two frames.

11.2.3 PERCEPTION

Perception is about *understanding the environment*, including object recognition and tracking. This can be achieved with *Deep Learning* which has seen rapid growth in recent years. It implements accurate object detection and tracking, using camera inputs. A CNN is a type of deep neural network widely used in object recognition tasks. A general CNN evaluation pipeline usually consists of the following layers: (1) the *convolution* layer which uses different filters to extract different features from the input image. Each filter contains a set of "learnable" parameters that are derived after the training stage; (2) the *activation* layer which decides whether to activate the target neuron or not; (3) the *pooling* layer which reduces the spatial size of the representation to reduce the number of parameters and consequently the computation in the network; and last, (4) the *fully connected* layer which connects all neurons to all activations in the previous layer. Once an object is identified, object tracking technology can be used to track nearby moving vehicles, as well as pedestrians crossing the road to ensure the current vehicle does not collide with moving entities.

11.2.4 DECISION MAKING

In the decision-making stage, action prediction and path planning mechanisms are combined to *generate an effective action plan* in real time. Indeed, the main challenge of autonomous driving planning is to ensure that autonomous vehicles travel safely in complex traffic environments. The decision-making unit generates predictions of nearby vehicles before deciding on an action plan based on these predictions. To predict the actions of other vehicles, one can generate a stochastic model of the reachable position sets of the other traffic participants and associate these reachable sets with probability distributions.

Planning the path of an autonomous vehicle in a dynamic environment is a complex problem, especially when the vehicle is required to use its full maneuvering capabilities. One approach is to search all possible paths and utilize a cost function to identify the best path. However, this requires an enormous amount of computational resources if the system is to be capable of delivering real-time navigation plans. To circumvent this requirement for computational complexity and provide effective real-time path planning, probabilistic planners can be utilized.

11.2.5 HD MAP CREATION AND MAINTENANCE

Traditional digital maps are usually generated from satellite imagery and have meter-level accuracy. Although this accuracy is sufficient for human drivers, autonomous vehicles demand maps with higher accuracy for lane-level information. Therefore, HD maps are needed for autonomous driving.

Just as with traditional digital maps, HD maps have many layers of information. At the bottom layer, instead of using satellite imagery, a grid map is generated by raw LiDAR data, with a grid granularity of about 5 × 5 cm. This grid basically records elevation and reflection information of the environment in each cell. As the autonomous vehicles are moving and collecting new LiDAR scans, they perform self-localization by performing a real time comparison of the new LiDAR scans against the grid map with initial position estimates provided by GNSS.

On top of the grid layer, there are several layers of semantic information. For instance, lane information is added to the grid map to allow autonomous vehicles determine whether they are on the correct lane when moving. On top of the lane information, traffic sign labels are added to notify the autonomous vehicles of the local speed limit, whether traffic lights are nearby, etc. This gives an additional layer of protection in case the sensors on the autonomous vehicles fail to catch the signs.

Traditional digital maps have a refresh cycle of 6–12 months. However, to make sure the HD maps contain the most up-to-date information, the refresh cycle for HD maps should be shortened to no more than one week. As a result, operating, generating, and maintaining HD maps can cost upward of millions of dollars per year for a mid-size city.

11.2.6   SYSTEM INTEGRATION

The above-mentioned components, such as the planning and control algorithms and the object recognition and tracking algorithms, have very different behavioral characteristics which call for different kinds of processors [3, 5].

HD maps, on the other hand, stress the memory. Therefore, it is imperative to design a computing hardware system which addresses these demands, all within limited computing resources and power budget. For instance, an early design of an autonomous driving computing system was equipped with an Intel® Xeon E5 processor and 4–8 Nvidia® K80 GPU accelerators, connected with a PCI-E bus. At its peak, the whole system, while capable of delivering 64.5 Tera Operations Per Second (TOPS), consumed about 3,000 W, consequently generating an enormous amount of heat. Also, at a cost of $30,000, the whole solution would be unaffordable (and unacceptable) to average consumers.

Recently, to tackle the computing power and cost problems, Tesla announced the development of its Full Self Driving (FSD) computing system, a 260 square millimeter piece of silicon, with 6 billion transistors, that Tesla claims offers 21 times the performance of the Nvidia chips it was using before.

11.3   ACHIEVING AFFORDABILITY AND RELIABILITY

Many major autonomous driving companies, such as Waymo, Baidu, and Uber, and several others are engaged in a competition to design and deploy the ultimate ubiquitous autonomous vehicle which can operate reliably and affordably, even in the most extreme environments. Yet, we have just seen that the cost for all sensors could be over $100,000, the cost for the computing system another $30,000, resulting in an extremely high cost for each vehicle: a demo autonomous vehicle can easily cost over $800,000. Further, beyond the unit cost, it is still unclear how the operational costs for HD map creation and maintenance will be covered.

In addition, even with the most advanced sensors, having autonomous vehicles co-exist with human-driven vehicles in complex traffic conditions remains a dicey proposition. As a result, unless we can significantly drop the costs of sensors, computing systems, and HD maps, as well as dramatically improve localization, perception, and decision-making algorithms in the next few years, autonomous driving will not be universally deployed.

Addressing these problems, a reliable autonomous LSEV has been developed by PerceptIn, for a total solution cost, excluding the chassis, under $10,000 and for scenarios with speed under 20 miles per hour, such as transportation services for university campuses, industrial parks, and areas with limited traffic. This approach starts with low speed to ensure safety, thus allowing immediate deployment. Then, with technology improvements and with the benefit of accumulated experience, high-speed scenarios will be envisioned, ultimately having the vehicle's performance equal that of a human driver in any driving scenario. The keys to enable affordability and reliability include using sensor fusion, modular design, and high-precision visual maps.

Figure 11.1: PerceptIn autonomous vehicle design.

11.3.1 SENSOR FUSION

Using only LiDAR for localization or perception is extremely expensive and may not be reliable. To achieve affordability and reliability, multiple affordable sensors, including cameras, Inertial Measure Units (IMU), GNSS receivers, wheel encoders, radars, and sonars,can be used to synergistically fuse their data. Not only do these sensors each have their own characteristics, drawbacks, and advantages but they complement each other such that when one fails or otherwise malfunctions, others can immediately take over to ensure system reliability. With this sensor fusion approach, sensor costs are limited to under $2,000.

The localization sub-system relies on GNSS receivers to provide an initial localization with sub-meter-level accuracy. Visual Inertial Odometry (VIO) relies on IMUs and cameras to further improve the localization accuracy down to the decimeter-level. In addition, wheel encoders can be used to track the vehicles' movements in case of GNSS receivers and VIO failures. Note that VIO deduces position changes by fusing IMU and visual data. However, when a sudden motion is applied to the vehicle, such as a sharp turn, it is possible that VIO will be less accurate due to the lack of overlapping regions between two consecutive visual frames as well as accumulation of IMU errors.

Therefore, to achieve reliable localization results, the *DragonFly* system developed by PerceptIn integrates multiple cameras into one hardware module, such that a pair of cameras faces the front of the vehicle and another pair of cameras faces the rear [4, 6, 7, 8, 9]. The active perception sub-system seeks to assist the vehicle in understanding its environment. Based on this understanding and a combination of *DragonFly* and of millimeter-wave radars to detect and track static or moving objects within a 100-m range, the vehicle can make action decisions to ensure a smooth and safe trip. The *DragonFly* module can capture spatial information, if deep-learning-based object recognition technology is applied, not only can objects including pedestrians and moving vehicles can be easily recognized, but the distance to these detected objects can be accurately pinpointed as well. In addition, millimeter-wave radars can also detect and track fast-moving objects and their distances under all weather conditions.

The passive perception sub-system aims to detect any immediate danger and acts as the last line of defense of the vehicle. It covers the near field, i.e., a range of 0–5 m around the vehicle. This is achieved by a combination of millimeter-wave radars and sonars. Radars are very good moving object detectors and sonars are very good static object detectors. Depending on the current vehicle speed, when something is detected within the near field, different policies are put into place to ensure the safety of the vehicle.

11.3.2 MODULAR DESIGN

In the recent past, designs of autonomous driving computing systems have tended to be costly but the PerceptIn design has demonstrated that affordable computing solutions are possible. This has been made possible by the application of modular design principles which push computing to the sensor end so as to reduce the computing demands on the main computing units. Indeed, the DragonFly module alone can generate image data at a rate of 400 MB/s. If all the sensor data was transferred to the main computing unit, it would require this computing unit to be extremely complex, with many consequences in terms of reliability, power, cost, etc.

PerceptIn's approach is more practical: it entails breaking the functional units into modules and having each module perform as much computing as possible. This makes for a reduction in the burden on the main computing system and a simplification in its design, with consequently higher reliability. More specifically, a GPU is embedded into the DragonFly module to extract features from the raw images. Then, only the extracted features are sent to the main computing unit, reducing the data transfer rate a thousand-fold. Applying the same design principles to the GNSS receiver sub-system and the radar sub-system reduces the cost of the whole computing system to less than $2,000.

11.3.3 HIGH-PRECISION VISUAL MAP

Creating and maintaining HD maps is another important component of deployment costs. Crowd-sourcing the data for creating HD maps has been proposed. However, this would require vehicles with LiDAR units, and we have already seen that LiDARs are extremely expensive and thus not ready for large-scale deployment. On the other hand, crowd-sourcing visual data is a very practical solution as many cars today are already equipped with cameras.

Hence, instead of building HD maps from scratch, PerceptIn's philosophy is to enhance existing digital maps with visual information to achieve decimeter-level accuracy. These are called High-Precision Visual Maps (HPVM). To effectively help with vehicle localization, HPVMs consists of multiple layers.

1. The bottom layer can be any of the existing digital maps, such as Open Street Map; this bottom layer has a resolution of about 1 m.

2. The second layer is the ground feature layer. It records the visual features from the road surfaces to improve mapping resolution to the decimeter level. The ground feature layer is particularly useful when in crowded city environments where the surroundings are filled with other vehicles and pedestrians.

3. The third layer is the spatial feature layer, which records the visual features from the environments; this provides more visual features compared to the ground feature layer. It also has a mapping resolution at the decimeter level. The spatial feature layer is particularly useful in less-crowded open environments such as the countryside.

4. The fourth layer is the semantic layer, which contains lane labels, traffic lights, traffic sign labels, etc. The semantic layer aids vehicles in making planning decisions such as routing.

11.4 DEPLOYING AUTONOMOUS LSEV FOR MOBILITY AS A SERVICE

In the previous section, we showed how to achieve affordability and reliability in building autonomous LSEVs. In this section we explain how autonomous LSEVs can be beneficial as part of the Mobility-as-a-Service (MaaS) ecosystem.

In today's MaaS ecosystem, ridesharing services such as Uber and Lyft usually cover a travel distance of more than 5 miles, whereas micro-mobility services such as Lime and Bird scooters usually cover less than 1 mile. The 1–5 miles travel distance actually accounts for 60% of total passenger miles traveled, but this travel distance is not well covered in the current MaaS ecosystem, either in the form of ridesharing or public transportation.

During the last few years of commercialization of autonomous LSEV technologies, we have discovered that many of the 1–5 miles trips actually take place in environments with limited traffic, such as university campuses, suburban areas, industrial parks, etc. In these areas, ridesharing services and public transportation services are usually either not available or too costly to maintain due to the high costs of drivers. Therefore, it is economically reasonable and technologically possible today to deploy autonomous LSEVs to handle 1–5 miles distance in the MaaS ecosystem.

Figure 11.2: PerceptIn autonomous shuttle operation in Japan.

In addition, since many of these 1–5 miles trips happen in environments with limited traffic and low speed limits, we do not have to mix the autonomous LSEV traffic, which has speeds less than 20 miles per hour, with the regular traffic, which may have speed as high as 60 miles per hour. It is therefore much safer to deploy autonomous LSEVs in these environments. With more and more success stories in deploying autonomous LSEVs, we believe we will soon witness large-scale adoption of autonomous LSEVs in the MaaS ecosystem to fulfill the needs of 1–5 miles transportation.

11.5 CONCLUSIONS

While there has been much progress over the past decade, it will probably be another decade or more before fully autonomous cars start taking to most roads and highways. In the meantime, a practical approach is to use low-speed autonomous vehicles in restricted settings. Then, as the relevant technology advances, the types of vehicles and deployments can expand, ultimately to include vehicles that can equal or surpass the performance of an expert human driver in any situation. PerceptIn has shown that that it is possible to build small, low-speed autonomous vehicles for much less than it costs to make a highway capable autonomous car. Not too far in the future, it might be possible for such clean-energy autonomous shuttles to be carrying passengers in city centers, such as downtown Manhattan, where the average speed of traffic now is only 7 miles per hour Such a fleet would significantly reduce the cost to riders, improve traffic conditions, enhance safety, and improve air quality to boot. Tackling autonomous driving on world's highways can come later.

11.6 REFERENCES

[1]   Liu, S. 2020. *Engineering Autonomous Vehicles and Robots: The DragonFly Modular-based Approach.* John Wiley and Sons. DOI: 10.1002/9781119570516. 203

[2]   Liu, S. and Gaudiot, J.L. 2020. Autonomous vehicles lite self-driving technologies should start small, go slow. *IEEE Spectrum*, 57(3), pp. 36–49. DOI: 10.1109/MSPEC.2020.9014458. 203

[3]   Tang, J., Yu, R., Liu, S., and Gaudiot, J.L. 2020. A container based edge offloading framework for autonomous driving. *IEEE Access*, 8, pp. 33713–33726. DOI: 10.1109/ACCESS.2020.2973457. 207

[4]   Tang, J., Liu, S., Liu, L., Yu, B., and Shi, W. 2020. LoPECS: A low-power edge computing system for real-time autonomous driving services. *IEEE Access*, 8, pp. 30467–30479. DOI: 10.1109/ACCESS.2020.2970728. 209

[5]   Liu, S., Liu, L., Tang, J., Yu, B., Wang, Y., and Shi, W. 2019. Edge computing for autonomous driving: Opportunities and challenges. *Proceedings of the IEEE*, 107(8), pp. 1697–1716. DOI: 10.1109/JPROC.2019.2915983. 207

[6]   Liu, Q., Qin, S., Yu, B., Tang, J., and Liu, S. 2020. π-BA: Bundle adjustment hardware accelerator based on distribution of 3D-point observations. *IEEE Transactions on Computers*. DOI: 10.1109/TC.2020.2984611. 209

[7]   Fang, W., Zhang, Y., Yu, B., and Liu, S., 2018. DragonFly+: FPGA-based quad-camera visual SLAM system for autonomous vehicles. *Proceedings IEEE HotChips*, p. 1. 209

[8]   Fang, W., Zhang, Y., Yu, B., and Liu, S. 2017. December. FPGA-based ORB feature extraction for real-time visual SLAM. In *2017 International Conference on Field Programmable Technology (ICFPT)* (pp. 275–278). IEEE. DOI: 10.1109/FPT.2017.8280159. 209

[9]   Yu, B., Hu, W., Xu, L., Tang, J., Liu, S., and Zhu, Y. 2020. Building the computing system for autonomous micromobility vehicles: Designconstraints and architectural optimizations. In *2020 53rd Annual IEEE/ACM International Symposium on Microarchitecture (MICRO)*, IEEE. 209

Author Biographies

Dr. Shaoshan Liu is the Founder and CEO of PerceptIn, an autonomous driving technology company. Since founding, PerceptIn has attracted over 12 million USD of funding of from top-notch venture capital firms, including Walden International, Matrix Partners, and Samsung Ventures. Before founding PerceptIn, Dr. Shaoshan Liu had over 10 years of experience at leading R&D institutes, including Baidu USA, LinkedIn, Microsoft, Microsoft Research, INRIA, Intel Research, and Broadcom. Dr. Shaoshan Liu received a Ph.D. in Computer Engineering from the University of California, Irvine. He has published over 60 high-quality research papers and holds over 150 U.S. international patents on robotics and autonomous driving and is also the lead author of the best-selling textbooks *Creating Autonomous Vehicle Systems* and *Engineering Autonomous Vehicles and Robots*. Dr. Shaoshan Liu is a senior member of IEEE, a distinguished speaker of the IEEE Computer Society, and a distinguished speaker of the ACM. Dr. Shaoshan Liu is also the founder of the IEEE Special Technical Community on Autonomous Driving Technologies. E-mail: shaoshan.liu@perceptin.io.

Dr. Liyun Li has more than 6 years of experience in autonomous driving software development. He is currently a principal engineer and manager at Xpeng Motors (NYSE: XPEV), where he leads the software development of Navigation Guided Pilot (NGP). Before joining Xpeng Motors, he served as a principal engineer at JD.com. He is one of the founding members of Baidu USA's autonomous driving team, where he has driven and led the effort of building core modules in Baidu's open-source autonomous driving system, including planning and prediction. Dr. Li has published two books in Autonomous Driving: *Creating Autonomous Vehicle Systems* (Morgan & Claypool Publishers) and *The First Technology Book in Autonomous Driving* (Publishing House of Electronics Industry (PHEI)). He is also the inventor of more than 20 international patents in autonomous driving. Dr. Li received his Ph.D. in Computer Science from New York University and Bachelor's degree in Electronic Engineering from Tsinghua University.

Dr. Jie Tang is currently an associate professor in the School of Computer Science and Engineering of South China University of Technology, Guangzhou, China. Before joining SCUT, Dr. Tang was a postdoctoral researcher at the University of California, Riverside and Clarkson University from December 2013 to August 2015. She received a B.E. from the University of Defense Technology in 2006, and a Ph.D. from the Beijing Institute of Technology in 2012, both in Computer Science. From 2009–2011, she was a visiting researcher at the PArallel Systems and Computer Architecture Lab (PASCAL) at the University of California, Irvine, USA. E-mail: cstangjie@scut.edu.cn.

Dr. Shuang Wu is currently a scientist at Yitu Technology. Previously he was a senior research scientist at Baidu's AI lab in Sunnyvale, CA and a senior architect at Baidu USDC. He earned his Ph.D. in Physics from University of Southern California, and was a postdoctoral researcher at UCLA. He has conducted research in computer and biological vision, applied machine learning in industry for computational advertisement, and speech recognition. He has published in conferences such as NIPS and ICML.

Dr. Jean-Luc Gaudiot received the Diplôme d'Ingénieur from ESIEE, Paris, France, in 1976 and M.S. and Ph.D. degrees in Computer Science from UCLA in 1977 and 1982, respectively. He is currently a professor in the Electrical Engineering and Computer Science Department at UC Irvine (UCI). Prior to joining UCI in 2002, he was Professor of Electrical Engineering at the University of Southern California since 1982. His research interests include multithreaded architectures, fault-tolerant multiprocessors, and implementation of reconfigurable architectures. He has published over 250 journal and conference papers. His research has been sponsored by NSF, DoE, and DARPA, as well as a number of industrial companies. He has served the community in various positions and was elected to the presidency of the IEEE Computer Society in 2017. E-mail: gaudiot@uci.edu.